W0081692

A Parent's Guide
to Crystals

The Group of 5 Crystals Series

Crystals and Stones: A Complete Guide to Their Healing Properties

The Eight Crystal Alliances: The Influence of Stones on the Personality

Publications by Paume de Saint-Germain Publishing

Jyoti for Kids: A Meditative Technique of Purification by the Light by Simhananda

The Lion's Roar: The Master from Montreal by Klaire D. Roy

New Tantrism by Klaire D. Roy

Tantric Training in the Age of Ray 7 by Klaire D. Roy

New Tantrism, Introductory Themes by Klaire D. Roy

Conclave of the Cryptic 7, Volume I by Klaire D. Roy

The Spiritual Science of Essential Yoga: Techniques of Meditation, Mantrams, and Invocations, Volume I by Sri Adi Dadi, compiled by Martine G. Fortier

Publications by North Atlantic Books

The Book of Stones: Who They Are and What They Teach by Robert Simmons and Naisha Ahsian

Stones of the New Consciousness: Healing, Awakening and Cocreating with Crystals, Minerals and Gems by Robert Simmons

The Mysterious Story of X7: Exploring the Spiritual Nature of Matter by Anonymous (Anne K. Edwards), introduction by Robert Sardello, foreword by Sir George Trevelyan

Steps on the Stone Path: Working with Crystals and Minerals as a Spiritual Practice by Robert Sardello

A Parent's Guide to Crystals

Gemstones to Support Your Child's Health and Happiness

The Group of 5

Paume de Saint-Germain Publishing
Montreal, Quebec, Canada

North Atlantic Books
Berkeley, California

Copyright © 2012 by the Group of 5. All rights reserved. No portion of this book, except for brief review, may be reproduced, stored in a retrieval system, or transmitted in any form or by any means—electronic, mechanical, photocopying, recording, or otherwise—without the written permission of the publisher. For information contact North Atlantic Books.

Published by

North Atlantic Books
P.O. Box 12327
Berkeley, California 94712

and

Paume de Saint-Germain Publishing
235 Rene Levesque Blvd. East, Suite 310
Montreal, Quebec, H2X 1N8 Canada

Translated from the original French by Jill Capri
Cover design by Susan Quasha
Book design by Lucie Robitaille and Lucie Létourneau
Photography by B. Simhananda and Gaëtan A. Brouillard, © Lux Æterna
Printed in the United States of America

A Parent's Guide to Crystals: Gemstones to Support Your Child's Health and Happiness is sponsored by the Society for the Study of Native Arts and Sciences, a nonprofit educational corporation whose goals are to develop an educational and cross-cultural perspective linking various scientific, social, and artistic fields; to nurture a holistic view of arts, sciences, humanities, and healing; and to publish and distribute literature on the relationship of mind, body, and nature.

North Atlantic Books' publications are available through most bookstores. For further information, visit our website at www.northatlanticbooks.com or call 800-733-3000.

Library of Congress Cataloging-in-Publication Data

Magie des pierres et des cristaux. English.
A parent's guide to crystals : gemstones to support your child's health and happiness / [translated from the original French by Jill Capri].
       p. cm. — (The group of 5
  Includes index.
  ISBN-13: 978-1-58394-496-7
  ISBN-10: 1-58394-496-6
1. Crystals—Health aspects. 2. Children—Health and hygiene. I. Title.
RZ560.M3413 2012
133'.25538—dc23

2012011782

1 2 3 4 5 6 7 8 9 United 17 16 15 14 13 12

To Dadi, our source of inspiration

ACKNOWLEDGMENTS

We also thank:

Our children, those founts of joy who enrich our lives with the magic of their presence.

The children, big and small, who have come to us for help and taught us invaluable lessons.

All the children around the world who are reaching out; they are our living sources of inspiration.

All the parents we know, and those we don't, who provide vital support and encouragement for their children. As we wrote this book, you were always in our thoughts.

CONTENTS

FOREWORD

Who knows whether the kinship between humans, stones, and crystals is not simply a memory left over from our early childhood? Even though the impact of our first encounter may have faded over time, most of us forged close relationships with the mineral realm at a very young age. Let's take a short stroll down memory lane. On the way home from school or hiking along a trail through the woods or along the shore, haven't we all been attracted by these curiously shaped shimmering objects, little treasures that we unobtrusively slipped into our pockets, secretly bringing home new friends charged with magical powers? Were we then able to catch a glimpse of the mystery of life in these small pebbles? Did we invest them with supernatural powers visible only to ourselves?

Is it possible that children are the only ones able to perceive these hidden secrets? Yet entire civilizations have been captivated by the properties of stones and crystals. Since the beginning of time, stones have fascinated us and shared their virtually magical qualities. Just think of the magnificent gems adorning the crowns of kings and queens, or the bejeweled swords wielded by history's greatest warriors. And let's not forget the never-ending search on the philosopher's stone. Traces of this forgotten knowledge can be found in all the great civilizations of the past.

Once again today, the world is awakening to this ancient mystery. We are at the dawn of a period of renewal, a period when science is redefining itself and when Mother Earth reveals to us a host of new minerals each day. As the quartz watches on wrists from Juneau to Jakarta so clearly illustrate, we are well and truly in the age of quartz.

The study of stones and crystals and their properties can be an enigma to many adults who tend to rely strictly on the rational side of their nature. Yet children are still enthralled and enchanted by this universe. Responsive to their environment and having not yet raised any mental barriers, children are innately sensitive to the energetic qualities of stones. That's why it makes sense to let them choose their own from the many that could correspond to their needs. Their intuition is their best guide. Interestingly, pastel colors are considered more appropriate for the straightforward energy of the young. For instance, rose quartz is a universal panacea that offers gentle protection. Since children react rapidly and effectively to stones' energy, they can benefit from carrying them in their pockets, or wearing them as a bracelet or necklace. As well, elixirs are easy to use and an efficient aid for parents seeking to help a child in need of particular treatment or therapy.

Of course, there is no doubt that the best therapy for children is caring parents whose love and attention are the most precious therapeutic tools. They are the basic ingredients of a universal and all-powerful remedy that is essential for all potential cures, which are founded on love and faith. Belief in an eventual cure is thus an integral part of the healing process. To children, parents are the supreme authority in all things.

Scientists too often criticize what is called the placebo effect, which sometimes provides the same benefits as medication and is the bane of the major pharmaceutical laboratories. Far from having a negligible or excessive impact, the placebo is a natural miracle cure whose effectiveness has been scientifically demonstrated on more than one occasion. What's more, another free formula, "the elixir of life and of healing," a remedy as old as human life itself, can on its own alleviate close to fifty percent of the diseases afflicting the human race. Any act to eliminate a "dis-ease" must be carried out with the conviction and faith that the problem can be resolved, as demonstrated by children who see their mothers and fathers as the best doctors in the world.

In our ever-more virtual world, there is an urgent need to return to our roots, to the earth that sustains us, to this immense spacecraft conveying us through the cosmos. An approach using unaltered minerals constitutes the perfect point of contact to support us in adulthood and maintain the purity and vitality of children. Stones not only serve us, they also contribute to the well-being of our brothers and sisters the animals, and our cousins the plants. A home embellished with specific stones confers more benefits than we can possibly imagine. Everyone knows how comfortable and cozy dwellings made of natural wood and stone can be. Imagine how decorating a child's room with the appropriate stones and crystals can create a climate conducive to creativity and study or simply a soothing atmosphere. Morpheus, the god of dreams in Greek mythology, is said to adore stones and crystals, to watch over children and bring them sweet dreams.

Contact with stones and crystals helps children appreciate the magic and simplicity of the little things that make up their day-to-day lives. They gradually acquire self-confidence and learn to respect the world around them. Parents see a new relationship develop between their children and their surroundings. A natural and trusting bond is effortlessly formed in this new triangle created by the stone, the child, and the parent.

A teacher who shares my interest in the mineral world gave me the opportunity to share my love of stones and meditation with a kindergarten class. I was overwhelmed by the response I received from these children. Because their minds are still open, they thought it was perfectly natural for stones to have their own properties and were eager to learn more. At the end of the class, one child assured me that her stone had spoken to her and silently answered her questions.

The "intelligence" of the mineral kingdom is astonishing. Each mineral is made up of a particular composition of chemical elements, and each chemical composition gives it a specific shape. The number of shapes and colors, the sheer artistry that is released from the bowels of the earth or even from the cosmos itself is truly amazing. God's "pharmacy" manifests itself in the wisdom of minerals, and the Great Architect has gifted us stones and crystals of the most extraordinary shapes. It is up to us to forge an alliance with these most basic yet supportive and powerful partners that are always prepared to serve us.

Gaëtan A. Brouillard, MD
Director, Clinique Médicale de Santé Globale Brouillard, Laval, Quebec; consulting physician, Maisonneuve Rosemont Hospital, Montreal

INTRODUCTION

The Group of 5 is pleased to present *A Parent's Guide to Crystals: Gemstones to Support Your Child's Health and Happiness,* an indispensable reference guide jam-packed with information that will help your children sail tranquilly through the sometimes troubled waters of childhood.

Children adore handling stones and crystals, observing them, examining them as if they somehow unconsciously know that each of these jewels holds a secret, a hidden treasure that could be revealed to them. When they work or play with a stone, it almost instantly and unconditionally transmits its strength, soothing properties, warmth, and knowledge. It can work a miracle in the blink of an eye and instill in children the joy of living, confidence in themselves, and good health.

Rose quartz, for instance, alleviates and calms anxiety, soothes wounded hearts, and encourages sound, restful sleep. Some stones have properties that will help smooth a child's path at school. They provide support for study, stimulate concentration, and positively influence behavior. They can boost self-esteem and powers of expression, reducing the fear of ridicule or being judged. Creating a protective aura, they also make children less vulnerable to the negative influences of some groups.

On a more physical level, some stones are also invaluable for teething, fever, stomachaches, nosebleeds, insect bites, and so on.

The world of children is tightly intertwined with our own, bringing its share of worries, problems, and joys. Although The Group of 5 wrote this book to help children, it is also ideal for parents, as the solutions it recommends are relevant to their problems as well. Needless to say, an adult's toothache can be alleviated by the same stones as those suggested for children.

This book is essential bedside reading for all mothers and fathers and a genuine treasure trove for any household. Thanks to support from the remarkable mineral world, parents will be armed and able to care for, comfort, and assist their children as they face a broad range of physical and psychological challenges.

Klaire D. Roy
Director of the Medicine Buddha Mandala Institute and The Group of 5

❖ ❖ ❖

The Proper Use of Stones, Crystals, and Elixirs

Factors Influencing Stone and Crystal Prices

Many factors influence the price of a stone or crystal. First, the authenticity of the piece plays an important role in the use we make of it. An artificial turquoise, created in a laboratory, will not be as efficient as a real turquoise. So be wary of imitations; although certainly less expensive, they will also be less effective.

The purity of the piece is an equally important factor to consider when purchasing a stone or crystal. The more crystalline it is, the greater its capacity to convey light. It will transmit its therapeutic particularities in less time and will increase our chances of success. The rarity of a stone or crystal directly influences its worth, hence its price. So a crystal from Tibet will be more expensive than a crystal from Arkansas. The originality of a piece and the extent of its beauty are often considered criteria of rarity, which increase the value of a stone or crystal.

Dimensions of a Stone or Crystal

The choice of the dimensions of a stone or crystal depends on the use we want to make of it. In therapy, we tend to choose smaller pieces that will be comfortable for the body. In daily life, certain pieces are ideally worn as jewelry. The dimensions and location (on the neck, finger, directly on the skin, or in a pocket) have a direct impact on the desired effect at the site we want it to be produced. As a decorative piece, or to harmonize and increase the energy and light of a given area, it is advisable to favor more imposing pieces like geodes or quartz clusters, which can be placed in a room, such as an office, or any other location.

Care of Stones and Crystals

Stones and crystals are precious friends; owing to their high sensitivity, proper care requires a particular vigilance that necessitates specific treatments. Ignoring this rule can lead to the destruction of these companions. When in doubt, it is better to refrain and wait to be certain as to the approach to be used. However, we have found two methods that are efficient and without danger: incense and an amethyst or quartz cluster.

For incense, just pass it over the stone or crystal for a few minutes. Reciting a mantra may increase the purifying effects. The choice of a mantra is a personal one, but "O" has proved to be as potent as a more elaborate mantra.

The use of a quartz or amethyst cluster is easy: it is simply a matter of letting the piece rest on the cluster for a few hours. The quartz or amethyst cluster will clean and charge the piece by injecting it with a supplementary energy dose that will render it more effective. Several other cleaning methods are also available. However, we want to reiterate that vigilance is important, since a bad choice can be fatal to your companion. Do not hesitate to consult books or to contact a specialist for advice on the best method to use for the stone or crystal you want to clean.

Best-known Cleaning Methods

- sunlight, ideal for all quartz
- moonlight, preferred by kyanite
- fresh cold water, adored by the majority of stones and crystals
- salt, to be used with caution
- earth, in certain cases, particularly if a crystal is very damaged
- snow, as a replacement for cold water

Best-known Recharging Methods

- quartz cluster
- sunlight
- moonlight
- cold water
- incense
- mantras

Methods of cleaning and recharging may differ, and one may not necessarily include the other. For example, salt may clean but not recharge a stone.

Stone and Crystal Elixirs

Since not everyone can afford to buy a ruby or a sapphire because of their price or rarity, stone elixirs are an excellent alternative. Moreover, children react rapidly to elixirs, which often act faster than actually wearing the stone itself. This is because elixirs are made from the highest quality stones under conditions that allow them to impart their strongest properties at all times, and so they are immediately absorbed.

Using Elixirs

Stone elixirs can be taken orally, in the form of three to five drops under the tongue or in water, three to four times a day as needed. They can also be applied directly on the skin, preferably on the sole of the foot or the palm of the hand. About eight drops of an elixir can also be effective when added to bathwater.

Elixir, Stone, or Crystal?

Elixirs are fast-acting. Since they are in liquid form, they are limited in quantity but definitely not in effectiveness. High-quality elixirs retain their properties over several years even once the bottle has been opened. Another advantage is that they are relatively inexpensive yet have all the properties of the stone. They are easy to carry around with you, and if you lose your bottle, you'll have no trouble replacing it at a reasonable cost.

Stones and crystals remain companions for life. Depending on how they are used, they will tirelessly serve their owner time and time again. Wearing a beautiful stone is always a pleasure, but it's important to remember that its effectiveness is enhanced when it touches your skin.

Stone or Crystal Therapy

If you'd like to treat yourself to a brief stone or crystal therapy session at home, here's an easy and reliable way to do it:

- Choose the stones or crystals you'd like to work with.
- Find a quiet place to lie down.
- Place the stones and crystals nearby.
- Lie on your back.
- Place the stones or crystals on your body, starting at the root chakra and ending at the crown chakra.
- Remain in this position for about twenty minutes, with or without music.
- When the time is up, remove the stones one at a time, beginning with the stone on the crown chakra.
- Continue to relax in this position for about five to ten minutes before getting up.
- Clean the stones and crystals using the appropriate method.

Placing stones and crystals on the body

To optimize the effectiveness of stones and crystals, you should know the most appropriate place to put them, which is determined by the pain or disorder to be treated. You can place the stone directly on the skin, over the affected organ or on the tender area. In this case, the stone or crystal shouldn't be left in place more than 20 minutes at a time.

The following section explains the correspondences between chakras, stones and crystals. Recognizing and respecting these correspondences will help you more effectively alleviate disorders associated with certain parts of the body.

Since each chakra resonates with a color, the stone or crystal you select should be compatible with this color. The following instructions will help you place the stones or crystals in harmony with your body.

For the base or root chakra (1st chakra)

Colors: red, black

Examples of stones: black tourmaline, fire opal, garnet, hematite, red agate, red jasper, red tiger eye, ruby, zoisite

For the *hara* or sacral chakra (2nd chakra)

Color: orange

Examples of stones and minerals: carnelian, copper, coral, dark amber, dark citrine, dark yellow topaz, orange calcite, orange sapphire

For the solar plexus chakra (3rd chakra)

Colors: yellow, green

Examples of stones: amber, ametrine, chlorite quartz, epidote quartz, green calcite, green fluorite, jade, malachite, rutilated quartz, smoky quartz, sunstone, tiger eye, yellow calcite, yellow sapphire, yellow topaz, yellow tourmaline

For the heart chakra (4th chakra)

Colors: green, pink

Examples of stones: amazonite, aventurine, chrysoprase, green calcite, green garnet, green jasper, jade, kunzite, malachite, moldavite, moss agate, pink danburite, pink fluorite, rhodochrosite, rhodonite, rose quartz

For the throat chakra (5th chakra)

Color: blue

Examples of stones: aqua aura quartz, aquamarine, azurite, blue lace agate, blue fluorite, blue sapphire, chrysocolla, kyanite, lapis lazuli, larimar, tanzanite, turquoise

For the third-eye chakra (6th chakra)

Colors: indigo, purple

Examples of stones: amethyst, azurite, charoite, iolite, lepidolite, purple fluorite, sugilite, tanzanite

For the crown chakra (7th chakra)

Colors: clear, golden, violet

Examples of stones: ametrine, clear fluorite, clear quartz, danburite, diamond, Herkimer quartz, pearl, selenite, white sapphire, white topaz

◆ ◆ ◆

Part 1: Early Childhood

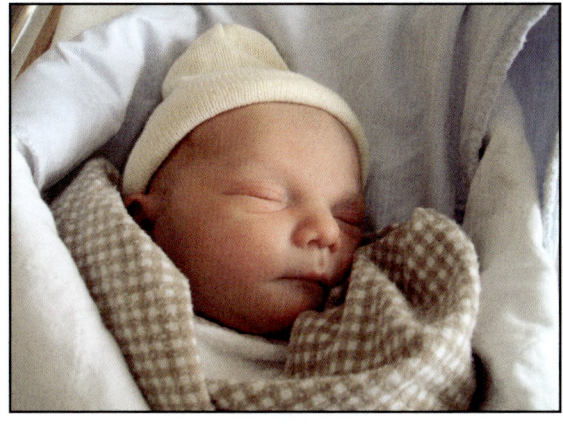
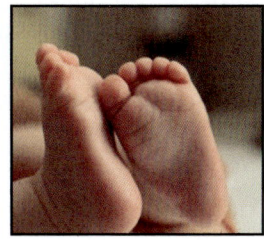

The First Months
with Your New Baby

Lisa C. Bergeron

The first months of a baby's life are incredibly challenging for the new mother's emotional body. Although there has been a physical separation between mother and child, they remain closely intertwined. The symbiosis is such that the mother may find it difficult to detach herself from her child's emotions, while at the same time she has to cope with her own emotional roller coaster triggered by dramatic hormonal changes. Day in, day out, new mothers have to adjust to scores of new and unexpected situations. They can often feel lost and inadequate when faced with the task of having to take care of two distressed emotional bodies on their own.

Furthermore, even though they know that all new mothers face similar challenges, they frequently feel alone. It is therefore essential that they receive the support they need. The mineral world, replete with knowledge and wisdom, can offer them a wealth of unsuspected resources. These stones and crystals can be strong allies as they learn the art of motherhood.

SUGGESTED STONES FOR MOTHERS

Carnelian

Carnelian represents the energy of Mother Earth and the wisdom of the Ancients. Containing the energy of fire, it reestablishes and sustains the energy of the physical body. It banishes negative emotions, including depression and fear. In the weeks after birth, it offers courage and physical support, and it regenerates and revitalizes the physicoenergetic body, particularly the reproductive organs that have been subjected to numerous shocks and changes. It improves the absorption of vitamins and nutrients, helps stop bleeding after birth, and stimulates the production of red blood cells. Carnelian also promotes scarring, the retraction of the uterus, and eases pain and cramps. For breast-feeding mothers, it stimulates milk production, as do chalcedony, chiastolite, and turquoise.

Lepidolite

Like tourmaline, lepidolite contains lithium, which relaxes the nervous system. A stone that invites tranquility, it stabilizes the emotional body and calms thoughts, freeing us from stress and anxiety. This marvelous stone is especially effective at the most difficult periods of our lives. It enables us to see challenges as learning opportunities and to appreciate the present moment for what it is. It also teaches us to listen to others with patience and compassion. Lepidolite helps new mothers more sensibly cope with the various situations they encounter.

Malachite

As the stone representing the Divine Womb, malachite communicates inner strength that encourages new mothers to trust their own instincts and take their rightful place within the family. A stone that has always been associated with beauty and femininity, it helps them remain true to themselves as they navigate the ongoing turbulence of their new lives. It fosters the expression of desires and feelings as well as releases inhibitions. This exceptional stone effectively calms the solar plexus and its emotional turmoil and is beneficial in healing the reproductive organs after birth.

Pink Tourmaline

Pink tourmaline awakens in us the feeling that life can offer us all we need. It strengthens the heart chakra by soothing the emotional body, calms the nervous system, and casts out depression and obsessive tendencies. Like ruby, it inspires courage and spurs us to protect those we love. It also encourages us to fulfill our commitments and duties when they are inspired by love. A strong ally of the emotions, pink tourmaline helps prevent new mothers from feeling invaded or imprisoned by their baby's cries, making sure they remain loving and attentive. It also helps them feel loved and supported.

It is important to point out that stones used by the mother influence the child, since mother and child are energetically linked, and the child is often in his or her mother's arms. It goes without saying that the more serene and rested the mother, the more contented the child.

SUGGESTED STONES FOR FATHERS

During the baby's first weeks, mother and child are as one and live in an enigmatic world that is often hard for the father to access. New fathers frequently say that they are puzzled by the presence of this strange new being they have to learn to love. They too need support in playing their new role and finding their place in this new family dynamic.

Moonstone

Moonstone can be a strong support for fathers who tend to be impatient or even clumsy and feel that being a father doesn't come naturally to them. It inspires patience, awakens intuition, and brings gentleness and tolerance. As well as guiding us to the right action at the right time, moonstone balances the masculine and feminine polarities in men and women. It helps fathers to understand the emotions of both the mother and the child and to be more attentive.

Pyrite

Pyrite, a marvelous fire stone, overcomes inertia, increases energy levels, and helps combat fatigue. It inspires self-knowledge and self-realization, imparting energy to help accomplish difficult day-to-day tasks. Pyrite supports masculine energy, boosts self-confidence, encourages the flow of ideas, and improves organization. It also helps us leave old habits behind to make room for new ones. This stone will guide fathers to draw on their resources and develop their potential, revealing the true father within.

Sunstone

Sunstone will help fathers who lack self-confidence or feel left out of the often fusional relationship formed between mother and child. It encourages us to meet challenges, emitting an inner light that brings joy, warmth, and good humor. It opens the mind to new perspectives, inspires us to become aware of our own worth, and boosts our self-confidence. Sunstone can help eliminate the stress, fear, and anxiety inherent in this new stage of our lives and brings optimism to new fathers. Fathers who feel overwhelmed with new lifelong responsibilities will feel supported by this stone, which will encourage them to fearlessly take action and get involved, while still respecting their masculine limits.

SUGGESTED CRYSTALS FOR BABIES

Celestite, Rose Quartz

Some stones and crystals are ideal companions for this new treasure that suddenly finds itself ejected from the protective cocoon that had sheltered it for nine months. Celestite and rose quartz bear the energy of maternal love, ensuring the newborn feels loved and safe. Both generate a gentle, harmonious energy when placed in a room, which facilitates the child's adaptation to its new earthly home. Celestite promotes healing, balances yin and yang energies, and teaches confidence in divine wisdom. It stimulates contact with the angelic realms and communicates deep peace to the child. Rose quartz conveys the gift of universal love; it calms, reassures, and eases the emotional body and also strengthens the heart.

Future parents, be prepared and assemble your first-aid kit before your baby arrives. Not only will you then be ready to experience this new joy to the fullest, but you'll also feel sustained and supported in your new role as parents.

◆ ◆ ◆

A New Baby in the Family

Lisa C. Bergeron

The arrival of a new baby is a happy but overwhelming occasion for all members of the family. While your newborn appears to be delighted to have landed in a warm, loving family, your older child is wondering why he has suddenly lost his place in the sun. Is he jealous, throwing tantrums, being confrontational, and showing signs of aggression? Is he constantly demanding your attention? Don't worry; that's all perfectly natural. He simply has to understand that he's no longer the only child that you shower with love and attention. Nonetheless, there are ways that you can help relieve his emotional distress.

Suggested Stones and Crystal

Aventurine

Aventurine nourishes and supports those who feel unloved or who find it difficult to open up their hearts. It inspires a feeling of satisfaction with life, favors empathy and compassion, and increases tolerance toward others.

As well as benefiting the heart chakra, aventurine comforts and calms emotional stress when used with rose quartz. As its name indicates, aventurine encourages adventure and makes it easier to accept the challenge of the arrival of a new brother or sister. In addition, it regularizes growth from birth to age seven.

Celestite

Celestite, which is a particular favorite for supporting children, is a deep sky-blue colored stone that is frequently used during early childhood since it imparts the energy of maternal love. It dissolves painful emotions, replacing them with a vibration of love. It also alleviates insecurity and soothes the angry emotions a young child often feels when a new baby enters the family.

Celestite maintains equilibrium between the flow of yin and yang energies, helping the child experience and express these two polarities in perfect balance and establishing inner well-being. This gentle stone brings peace to the child, encouraging openness to this new situation. It helps improve relationships and fosters smoother and clearer communications. Once the child can more easily express his or her needs, joys, and sorrows, the parents can take more appropriate action. It is an ideal stone to help maintain a harmonious atmosphere during stressful and demanding periods of our lives.

Dioptase

Dioptase draws in a high vibration of love at all levels, particularly at the heart. It thus has the power to rid this chakra of any ills afflicting it. Dioptase can also change negative into positive. Furthermore, it clarifies and purifies thoughts and lessens the tendency to want to control others, an attitude often displayed by older children toward their younger siblings.

Dioptase benefits children who feel they have lost their rightful place by mitigating feelings of betrayal, grief, sadness, abandonment, and trauma. For adults, it is also an excellent stone for healing the emotional wounds of the inner child. Also, it helps heal suffering in a more immediate fashion and thus prevents the negative consequences of unresolved emotional trauma or conflict.

Smithsonite

The gentle smithsonite resonates with the energy of Kwan Yin, the goddess of compassion. A facet of the divine mother, Kwan Yin is often called on for her healing qualities and ability to dissolve interpersonal conflicts and create unity. She awakens the inner comprehension of the universal energy of love and the certainty of the Unity of the Whole.

An emotional healer, smithsonite lessens feelings of loneliness, releases grief, and comforts broken hearts. It calms anger and resentment and confers courage on those who feel unloved and are going through troubled times.

It inspires gentleness and compassion, encouraging us to support others. Smithsonite promotes friendship through inner harmony and harmony with others. It has the power to bring people together, creating unifying energy when each child wears one.

It usually takes an older child several months, even up to a year or two, to adapt to the arrival of a new baby, as each child is different. Be patient and support your child throughout this process. A judicious selection of crystals, elixirs, and stones, together with patience and love, is usually a winning formula.

◆ ◆ ◆

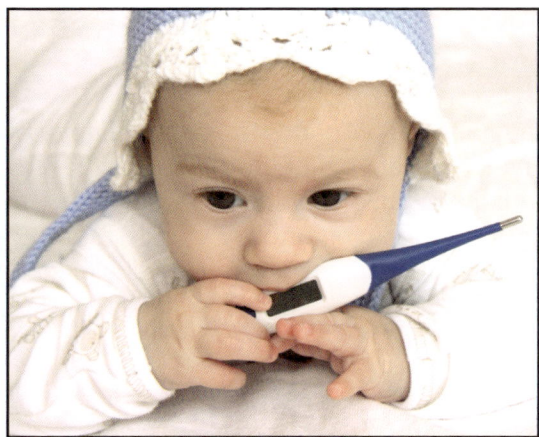

My Child Has a Fever

Ginette Tétreault

Children are said to have a fever when their body temperature rises above 100.4 degrees Fahrenheit (38 degrees Celsius). Usually not serious, a fever is simply the body's natural reaction to the presence of harmful agents in the organism. However, a doctor should be consulted if an infant under six months of age has a persistent fever.

Feverish children often exhibit the following symptoms:

- watery eyes
- flushed cheeks
- rapid breathing
- loss of appetite
- rapid heartbeat
- irritability

The healing powers of the mineral world effectively support children's immune systems. A sound mineral balance is the very basis of good physical health. Mother Earth offers us three stones with different qualities that have proved their effectiveness in reducing fever.

STONE PROMOTING SELF-HEALING

Amber

Known for its properties as a natural antibiotic, amber helps overcome illness by enabling the body to recover its balance and gradually heal itself. If the child is old enough, an amber necklace will alleviate fever caused by teething or other problems. Mothers should wear the stone first so that it can absorb their positive energy.

"SUN IN A STONE"

Sulfur

Like sponges, children absorb any negative energy in their family environment, which affects their immune system. In the same way, sulfur, which is yellow like the sun, absorbs negative energies and helps combat flare-ups of fever. Its mineral qualities are effective against viral infections, colds, and flu. In addition, thanks to its antibacterial and antiseptic properties, sulfur purifies the blood, intestines, and urine.

However, since it is composed of toxic agents, sulfur should be used with care, and only externally. If the skin becomes irritated when it is placed directly on the body, simply place a thin piece of cloth between it and the skin.

STONE PROMOTING PURIFICATION

Green Calcite

Green calcite emits emanations that raise a child's vibratory field above the illness. Its healing properties strongly reinforce the immune system, stimulate the thymus, relieve fever, and rid the body of bacterial infections. Because of its composition of calcium, manganese, and iron, it calms the nervous system.

Other Helpful Stones

SUGGESTED STONES

Hematite, Sodalite

The three tireless allies above—amber, sulfur, and green calcite—can also be supported by hematite and sodalite to relieve fever. We can count on these stones to lend assistance when we feel helpless in the face of our children's fever and discomfort. When they are also sustained by the energy of parental love, so essential to the healing process, our precious offspring will gradually find themselves on the road to recovery.

I'VE GOT A TUMMY ACHE

LISE DUSSAULT

Because they care about their children's health, parents are always on the lookout for anything unusual and pay particular attention to symptoms to determine how serious the disorder may be. So what should they do about a complaint as vague as a "tummy ache"? Is their child simply trying to divert their attention or manipulate them, or is it a symptom of a genuine ailment that requires the opinion of a health care professional? Certain signs and their degree of intensity clearly indicate when parents shouldn't hesitate to call on a physician. However, depending on the causes of these tummy aches, it isn't always necessary to visit the doctor. But even if the problem is psychological or emotional in origin, the actual stomachache is very real.

Supporting the Nervous System

A child's world today has its share of challenges and stress. A stomachache may indicate discomfort or tension that the child is unable to express in any other way. Visceral pain is said to be triggered or sustained by an overload of the autonomic nervous system; the body "communicates" the presence of an imbalance through the nervous system. Well-being may be restored in a variety of ways, including using stones and crystals to help enhance it. Luckily, certain stones and crystals can influence the nervous system.

SUGGESTED STONES AND CRYSTAL

Amazonite

Amazonite harmonizes the autonomic nervous system and internal organs. It dissolves worries, fears, and anger and balances mood swings.

Amethyst

Amethyst soothes nervous disorders, relieves pain in general, and inspires relaxation. It reduces fears of all kinds and calms violent emotions and anger, giving us the courage to face what the day may bring.

Selenite

Selenite has a soothing effect on the entire nervous system, promoting a feeling of physical well-being. It helps children remain centered and in harmony with the world around them. It also subtly protects from negative influences that may be directed toward them from their friends or from their own inner fears, such as the fear of ridicule and the fear of being unloved or unrecognized.

Communication Reduces Stress

As the nervous system may transmit a "disorder" throughout its network, good communication can help reestablish a healthy balance. Talking to children about what triggers their stress or anxiety helps them relax, brings relief, and steers them toward appropriate solutions. A number of stones and crystals are effective in promoting better communication. Interestingly, the stones and crystal described below not only favor personal interaction but also impart calm.

SUGGESTED STONES AND CRYSTAL

Aquamarine

Aquamarine encourages the expression of truth and inner feelings. It soothes, reduces stress, anxiety, and fears, and calms the mental faculties. It thus supports children by enabling them to express themselves and say what they think without fear of ridicule.

Blue Apatite

Blue apatite also promotes communication and self-expression. It helps overcome sorrow, anger, and lethargy while soothing the heart and emotional wounds.

Aqua Aura Quartz

Aqua aura quartz favors the sincere exchange of ideas. It also calms and balances the emotional body. In addition, it helps relieve stress and reduces aggressiveness and restlessness.

Self-confidence

Children need a healthy dose of self-confidence to help them face certain situations. The following stones and crystals can support them in building the self-confidence they need.

SUGGESTED STONES AND CRYSTALS

Amethyst or Selenite Cluster

Placing an amethyst or selenite cluster in a child's room promotes relaxation, tranquility, and restorative sleep. If your child likes to sleep with a night light, a blue bulb will encourage relaxation.

Calcite

Calcite inspires self-confidence and confers stability. It balances the emotional body, diminishes fears, and boosts individual energy.

Chrysoprase

Chrysoprase increases feelings of security and confidence. It develops our self-assurance and helps eliminate negative attitudes by focusing our attention on the positive aspects of events and situations.

Rose Quartz

Rose quartz also helps us acquire confidence. It imparts gentleness, tenderness, calm, and reassurance.

As parents, we have a multitude of roles to play to respond to our children's many needs. These needs emerge and change throughout the learning process and the challenges inherent in growing up. Many resources are available to parents to help them support and sustain their children at this time. However, even though listening closely to them is undoubtedly a solution that resolves many problems, we must remember that the most important remedy is to give them the love they need to heal and grow.

◆ ◆ ◆

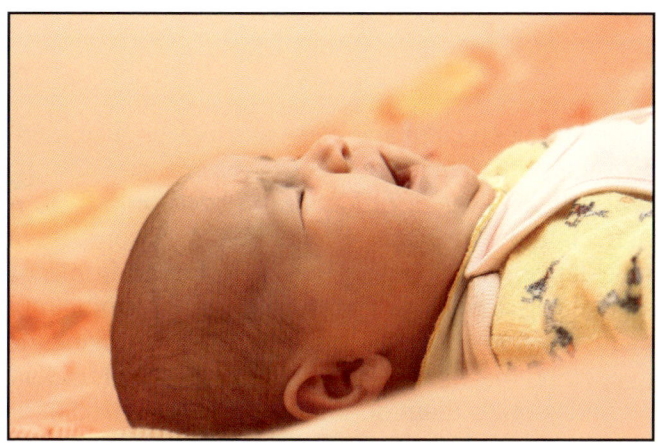

COLIC
LISA C. BERGERON

Bouts of colic are unfortunately part and parcel of the joyous experience of being new parents. Science today may have the answer to many questions, but there are still numerous mysteries that elude its grasp, and colic is one obvious example. Scientific research has been unable to explain exactly what this pain is and why it appears. A number of theories have been put forward, and various remedies are available that seem to work on a hit-or-miss basis. In one sense, the mystery surrounding colic may be its only compelling characteristic as it reminds us of the power of nature and forces us to take a fresh approach to various challenges. Who knows? The reason for a child's colic could be an opportunity for his or her parents to develop their intuition, exemplary patience, and unparalleled creativity.

Below are a few suggestions to help parents cope during this period.

SUGGESTED MINERAL

Magnesium
Magnesium has a calming and relaxing effect on both body and mind. Since it is a key element in relaxing tension as well as cramps or spasms in muscles, the stomach, the intestines, and the gall bladder, stones containing magnesium are ideal for colic.

SUGGESTED STONES

Magnesite

Magnesite conveys deep calm and alleviates tension, fear, and irritation. It balances magnesium deficiencies, eases spasms, and acts as a muscle relaxant. It also treats liver disorders and relieves intestinal, menstrual, and stomach cramps. Furthermore, magnesite encourages a positive attitude, acceptance, and self-love. Infants suffering from colic are, of course, experiencing stomach pain, but they are also exhibiting emotional distress because the treatment is often ineffective. Thanks to its high magnesium content, magnesite is a powerful emotional healer that helps infants relax and soothes them during this stressful period. It helps them feel secure and sustained by the Divine Being.

Malachite

By conveying compassion, malachite absorbs physical and emotional pain. It primarily soothes disorders of the solar plexus, cleanses the emotional and physical bodies, and protects from negative energies. It also alleviates anxiety, diminishes fear, transmits balance and harmony, and opens the consciousness to unconditional love. In addition, malachite reduces inflammation and acidity in tissues and acts as an antispasmodic agent to eliminate cramps. It also harmonizes DNA and the cellular structure and boosts the immune system.

Serpentine

A New Age stone and member of the jade family, serpentine directs healing energy to psychological and emotional imbalances. As well as attenuating mood swings, it reduces stress and nervousness, conferring a peaceful attitude in the midst of conflict. Serpentine transmits a calm, gentle vibration to the emotional body, thereby encouraging the release of the fear of change and adversity. It eases cramps and strengthens the stomach, intestines, bladder, and kidneys. It is a reenergizing stone that also promotes the absorption of calcium and magnesium.

Turquoise

A stone of protection, turquoise is possibly the oldest stone used by humans. Particularly prized for its regenerating qualities, it transmits strength and encourages optimal overall health. Turquoise aligns and balances the chakras and meridians. It calms the nervous system and also acts as an antispasmodic and analgesic agent. Turquoise neutralizes hyperacidity, soothes stomach problems, and imparts inner warmth throughout the body.

Stones and stone elixirs may be placed or applied directly on the infant's abdomen. By remaining attentive to the child's responses and observing the effect of the contact with the stone or elixir, parents can learn which stone is best for their baby. It's important to remember that as unpleasant and trying as this time may be, colic doesn't last forever.

◆ ◆ ◆

Teeth and Teething

Johanne Marier

All parents have sleepless nights when their babies are cutting their first teeth. This is often the first time in his or her life that a child encounters suffering. Our first experiences as supportive parents frequently reveal the lack of resources available to relieve their pain. In our search to bring them relief, we often turn to more traditional solutions. However, there are other options that have no side effects: stones are marvelous tools that can produce effective results and, we hope, awaken your interest in this fascinating field. The testimony of numerous parents who have been able to ease their child's teething pains thanks to the soothing support of stones and crystals confirms our belief in this gentle approach.

Teething

The appearance of their first teeth, which is often accompanied by tears, is an important step in all children's lives. They first start to get their baby teeth (incisors, eyeteeth, and molars) at about six months of age. In response, they may drool, suck their fingers, cry frequently, and stop eating.

SUGGESTED STONE

Amber

An amber pendant is beneficial for teething infants as long as the mother has worn it for a period of time beforehand to ensure that her energy is transmitted to the child. A piece of amber placed in a small cloth bag and attached to the child's clothing will also ease the pain. Amber elixir can also be useful in these cases; simply place a few drops close to the child's mouth or on the neck. All the stones suggested in this chapter are available in elixir form, which has proved to be very effective for children.

Tooth Development and Growth

After this initial pain, the child's teething process should develop normally. However, a poor diet, pacifiers, thumb sucking, and so on can all have an adverse impact on tooth development. Fortunately, some stones can help prevent deformed palates or dental malformation.

SUGGESTED STONES

Aventurine

Aventurine supports children during their development, regulating their growth from birth to age seven. More specifically, it has an anti-inflammatory action on sensitive gums. Aventurine contains silicon, an important mineral in the human body, which helps strengthen bones and absorb calcium.

Selenite

Known as a stone that is closely linked to the angels, selenite inspires deep calm and can encourage children to stop sucking their fingers. When chosen with care and worn as a pendant or simply placed in a child's room, selenite can provide significant support during teething.

Abscesses

An abscess is defined as a collection of pus that forms in a natural or incidental cavity of the body, which explains why they may be common during the teething process. An abscess creates an infection that is accompanied by painful swelling and inflammation.

SUGGESTED STONES AND CRYSTAL

Clear Quartz, Malachite

A number of stones have the necessary soothing properties to alleviate infection, inflammation, or pain. Some contain minerals such as copper that are effective in combating inflammation. Malachite, for example, which can relieve various disorders, reduces inflammation caused by dental abscesses. However, since this stone can tire as it absorbs, it can quickly lose its power. That's why it is recommended to use it with clear quartz, which can help it retain its strength.

Green Calcite, Fluorite, Magnetite

Green calcite controls infections, detoxifies the system, and reduces inflammation. Fluorite and magnetite ease teething pains and can relieve all intense pain.

Tooth Extraction

When children have to have a tooth pulled for the first time, they are often overwhelmed by fear of the unknown. This fear is exacerbated if the person accompanying them suffers from anxiety. But there's no cause for concern: two valuable allies are available to help alleviate this momentary discomfort.

SUGGESTED CRYSTALS

Amethyst

Amethyst is a must in any first-aid kit. It imparts gentle calm, helping parents and child face any new situation.

Celestite

Celestite can help dispel the pain of a tooth extraction. Children can take it with them into the dentist's office for support when their parents can't accompany them. In fact, celestite has the qualities needed to play a mother's role, bringing love, soothing fears, and dissolving anxiety.

Wearing Braces

A number of children have to wear braces or other orthodontic appliances. Many parents have told us about the various complications than can arise and have offered suggestions. In these circumstances, it's important to reassure children who are in pain or afraid of experiencing it.

SUGGESTED STONES

Selenite

Selenite is essential for regular even teeth and aids in the straightening process. It also neutralizes the toxins that may be released by the mercury in fillings.

Turquoise

Turquoise is a valuable partner and protector for children to take with them the first time they visit their orthodontist. It regulates the nervous system, has a protective effect, and maintains the mucous membranes in the mouth and throat.

Relieving Pain

SUGGESTED STONES

Aquamarine, Fluorite, Magnetite

A fluorite or magnetite pendant is an excellent way to ease the pain that intensifies in the hours after the braces have been attached. Also, simply placing the stones close to the teeth will alleviate the child's suffering.

Fluorite, magnetite, or aquamarine elixirs are recommended for intense pain. Regularly applying aquamarine elixir directly on the painful tooth can be very helpful. Placing a few drops of any of these elixirs under the tongue will sustain children and relieve their pain.

Strengthening Teeth

Stones can help strengthen teeth throughout the entire orthodontic treatment, which may sometimes last several years.

SUGGESTED STONES

Apatite

Apatite is an ideal stone for supporting children who wear an orthodontic appliance, since it helps them absorb calcium.

Calcite, Magnesite

Calcite and magnesite are essential during this period. Magnesite contains magnesium, a mineral salt present in great abundance in the body, and works in synergy with calcite, which contains calcium. These stones thus provide two indispensable elements for the formation of bones and teeth.

Ulcers

When braces rub against the gums, they can cause very painful ulcers or canker sores in the mouth. Orthodontists recommend placing wax between the gum and the tooth, but unfortunately this solution provides only temporary relief.

SUGGESTED STONE

Fluorite

Fluorite can produce spectacular results in helping to heal these ulcers. Applying fluorite elixir directly on the ulcer has also proved to be extremely effective.

Exhaustion

Some children may find orthodontic treatment exhausting: they have to wear braces, make frequent visits to various specialists where they are required to keep their mouths wide open for long periods of time, and so on. Children need to be protected against the negative energies they could absorb while under treatment.

SUGGESTED STONES

Garnet, Labradorite, Tiger Eye

Labradorite is both a protective and an energizing stone that can sustain children during this process. It can be worn as a pendant, as earrings, or taken in elixir form. Tiger eye is also an excellent protective stone. Lastly, garnet can help restore children's energy.

A Sure Favorite

SUGGESTED CRYSTAL

Rose Quartz

This chapter reviews various childhood dental problems and the teething process, recommending the appropriate stones, crystals, and elixirs that can support and sustain them at these difficult times. But we must not forget that rose quartz, lithotherapists' favorite stone, is also extremely useful in these cases. Loved by children, its gentle action promotes refreshing sleep, provides comfort, and has proved to be a strong support in the most troubled periods.

The above stones should be included in your kit to ease any temporary dental pain. We hope this kit will become an indispensable resource for you to provide the care and comfort your children need and bring a smile back to their faces.

◆ ◆ ◆

Banishing Fear
Philippe Soreau

Ogres, big bad wolves, and witches are all figures of fear that children love to hear about from a very early age. Yet the fun is over when they wake up crying in the night, are afraid of animals, or refuse to take a bath. Childhood fears often stem from threatening experiences they don't understand. How can we help them overcome these terrors?

Parents need to listen to their children and understand and comfort them so they can feel secure and comprehend their fears. Some stones and crystals that are used in lithotherapy and have soothing properties can also help children face up to their fears.

Small Children

Children up to age two are easily frightened by certain objects or noises such as the crash of thunder or even the roar of the vacuum cleaner. Strangers, unknown places, or being left alone can also upset or even terrify them.

Children from age two to six also have fears that they can't control. Being alone in the dark; having to pass through gloomy rooms or streets; the feeling of being chased, bitten, or devoured by monsters, ghosts, giants, ogres, or witches; being separated from their parents; or even simple things like visiting the doctor, being in contact with small or large animals, taking a bath, seeing lightning, or hearing strange noises in the city are all examples of childhood fears.

SUGGESTED STONES AND CRYSTALS

To complement the indispensable, reassuring, and comforting presence of parents, some stones can also support children and help them confront their fears. Celestite and rose quartz are particularly effective for very young children as they also reassure, comfort, and convey love.

Agate

Agate is a protective stone that eliminates all negativity. It provides calm and balance, dispels fear, and inspires a deep feeling of security. It also enhances concentration.

Amethyst

A powerful protective quartz, amethyst imparts peace and relaxation. It also helps calm strong emotions. Placed under the pillow, amethyst promotes restful sleep and prevents nightmares.

Aventurine

Aventurine bestows courage and soothes. It calms emotional stress and reduces stuttering. As well, it regulates a child's growth from birth to age seven. It may be used in conjunction with rose quartz for children.

Celestite

Celestite bears energy and maternal love, and as such is ideal for children. It calms agitated emotions and promotes deep peace, opening us up to new experiences. It also transmits confidence and divine wisdom.

Citrine

Citrine helps children rediscover their inner nature, that of joy and light. It inspires courage, assurance, and self-confidence. It also supports sensitive, vulnerable children and helps free them from oppressive influences.

Rose Quartz

As the stone of unconditional love and inner peace, rose quartz reassures, soothes, and provides encouragement at times of emotional shock. Placing rose quartz in a child's room releases this calm and gentleness, encouraging peaceful sleep.

Sugilite

Sugilite reduces the intensity of childhood fears by alleviating sadness and sorrow and promoting self-forgiveness.

Children Ages Six to Twelve and Teens

Children ages six to twelve have to confront the school environment and face all the challenges this involves. They form more and more relationships outside the immediate family circle and have the opportunity to discover and embark on new experiences. And new fears go hand in hand with these experiences: fear of school, failing, making and losing friends, sports, accidents, physical violence, kidnapping, fire, death of their parents, and death in general are just some examples.

Exposure to events that receive extensive media coverage can also be frightening for children. Witnessing a traumatic or terrifying event on the news or seeing scenes of terrorism, war, attacks, or hostage-taking in movies can also trigger childhood fears. Teenagers can also fall prey to fear, depending on their personality. They may, for instance, be troubled by speaking in public, by relationships with romantic interests, by sexual activity, or be concerned about their future.

SUGGESTED STONES AND CRYSTALS

Some stones and crystals can also help this age group successfully navigate this stage of their lives and its countless new experiences. They can help reduce their fears and promote emotional balance while fostering communication and self-expression. These stones and crystals provide protection and impart the courage and confidence these young people need.

Amethyst, Aventurine, Rose Quartz

Three powerful partners that we've already recommended for smaller children can also be suitable for older youth. Amethyst protects and conveys the calm needed to overcome the fear of speaking in public, for example. Rose quartz soothes, reassures, and encourages, while aventurine inspires strength and courage.

Carnelian

The stone of courage, carnelian inspires us to believe in ourselves and stand on our own two feet. It sweeps away negative emotions like fear and depression. It also raises our energy levels and clarifies thought. In addition, it conveys joy and happiness.

Citrine

Citrine is ideal for children since it helps them rediscover their inner nature, that of joy and light. It inspires courage, assurance, and self-confidence. It also supports sensitive, vulnerable children and helps free them from oppressive influences.

Herkimer Quartz

Herkimer quartz is a useful stone for children as it can eliminate negativity and attract joy and harmony. It has all the qualities of quartz, including the ability to promote the circulation of energy and purify negative energies. And since it also enhances the impact of other stones, Herkimer quartz may be used with practically all of them.

Kunzite

Kunzite is excellent for calming and balancing the emotions. It soothes anxiety and panic attacks and eliminates emotional dependence. It provides a protective shield against unwanted energies. Particularly useful for children on an emotional roller coaster, it can also support young people suffering from bipolar disorder thanks to its lithium content.

Lepidolite

When children experience frustration and anxiety caused by stress, lepidolite brings calm and grants independence. Its lithium content enables this stone to inspire tranquility and also to help combat epilepsy. Lepidolite protects us from outside influences and negative thoughts.

Orange Calcite

Orange calcite is an excellent stone for dispelling fears and instilling stability and self-confidence. It inspires joy and lightheartedness and helps children balance their emotions.

Sodalite

Sodalite is highly suitable for hypersensitive children. It helps transform their defensive or overly sensitive personality and enables them to leave their fears behind. As well, it reduces guilt and aids in eliminating the control mechanisms that prevent them from truly being themselves. It also assists in balancing the emotions and helps children express their feelings.

Tiger Eye

Tiger eye is a protective stone that reduces the negative influence of threatening situations and helps young people distance themselves from them, while at the same time protecting them from negative energies. It helps dissolve fear, banish anxiety, and stabilize the emotions. Tiger eye also transmits strength, willpower, courage, and self-confidence.

Countless stones and crystals are available that can help alleviate childhood fears. They can serve children as they progress through the various learning stages of their lives. These stones provide support, enabling them to assimilate and understand what they are learning and subsequently translate this knowledge into positive experiences for development and growth.

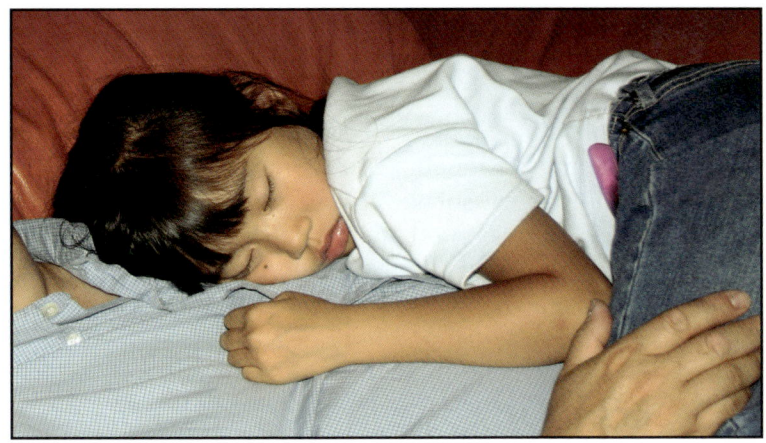

Nocturnal Fears

Kristiane Roy

Nighttime fears are a normal part of growing up in a world full of unknowns and mixed messages. Young children have vivid imaginations, and a pile of dirty laundry in a dark room can easily become a scary monster hovering by the bed. They may fear intruders, unexplained sounds, darkness, or supernatural creatures from a recent film; their fertile minds are full of magical thinking and fearful fantasies.

Irrational Fears

SUGGESTED STONES AND CRYSTAL

Charoite

The soft mauve color of charoite quickly soothes worry and stress in children, promoting peaceful sleep and sweet dreams. The mauve ray is also known to encourage profound emotional healing. Charoite's interspersed black markings act to penetrate deeper in the psyche in order to transmute fears, hence its moniker "The Stone of Transformation."

Rose Quartz

Rose quartz is known as the stone of unconditional love and inner peace. It confers gentleness and tenderness while calming worries and fears that cause insomnia. Rose quartz supports us in moments of emotional shock and inspires harmony in chaotic situations.

Sleeping Alone

Sleeping alone can create much anxiety in children. It is the only time when they are completely isolated from other people for long periods of time, and on top of this, in relative darkness. Parents may just be in the next room, but to a child they may feel miles away. In the Western world the security of a "family bed" is not a common practice, as it is in other cultures, so how can children feel the reassuring presence of Mom and Dad while staying in their own room?

SUGGESTED CRYSTAL

Celestite

Celestite is a pale blue crystalline stone with an affinity to the angelic realm. It radiates maternal love and promotes security during stressful times. It emits an energy of deep peace and is very soothing in a child's room.

Filtering the Outside World

As children get older they may be exposed to much troubling information and stimulation, from violent video games and disturbing news reports to movies that promote the darker side of human nature. While the game or movie may be over, the impact may remain long after the lights go out. Studies have shown that many Western parents are not aware of their older child's nighttime fears yet they are far more prevalent than we realize.[1] While certain fears are inevitable and part of normal childhood development, it is important to filter what seeps into their developing consciousness, so that normal fears do not become troubling concerns. So how can today's parent ease the nocturnal angst of their growing children?

SUGGESTED STONE

Kunzite

Kunzite balances and calms emotions and is even known to neutralize high anxiety and panic attacks. It allows one to be balanced and

---

[1] Gwen Dewar, *Nighttime Fears in Children: A Guide for the Science-minded Parent,* 2008. http://www.parentingscience.com/nighttime-fears.html.

focused amid chaos and distraction. It encourages wisdom and just action while allowing suppressed feelings to be expressed in a constructive manner. Kunzite facilitates maturity and self-respect, and provides a protective shield against undesired energies.

Fear of Death

Death is not a subject that is commonly talked about in our society, yet it is present everywhere, often seen in a most frightening and devastating light. Many children have to face their fear of death alone as many adults themselves are not prepared to face it realistically. Older children often fear the death of their parents, or worry about natural disasters, war, or accidents.[1] Death is as prevalent as birth, and it is important to demystify it for ourselves as well as for our children. Nightmares and night fears often circle around this subject, but parents can help by guiding their children to explore this great and mysterious passage that we all must one day pass through.

SUGGESTED CRYSTAL

Amethyst

Amethyst is known as the stone of courage. It offers protection, diminishes nightmares, and banishes negative thoughts. It promotes peaceful sleep and soothes the nervous system. Amethyst works to diminish many kinds of fears, including the fear of death. It offers assistance with the grieving process and supports those facing death or serious illness.

Dreaded Tomorrow

Certain difficulties and challenges may naturally evoke fear. Children may not always know how to talk about these worries, as they may not be equipped to face them accordingly. Family life may be heavy and unharmonious, a school bully may be lurking in the corridors, a strict teacher may demand more than they can handle, or they may have received a bad report card despite all efforts. All of this may come to life at night, as children attempt to face a dreaded tomorrow that has not yet arrived.

---

[1] Dewar, *Nighttime Fears in Children.*

SUGGESTED STONE

Aquamarine

Aquamarine incites us to speak the truth and to release emotions. It clears the mind and alleviates anxiety, fears, and morose thoughts. It lends courage and encourages perseverance during difficult times.

Suggestions:

- Place a stone under your child's pillow at night. Quartz such as amethyst, rose quartz, aquamarine, kunzite, or celestite is best as it irradiates through cloth. It is prudent to clean the stone every morning with cold water.
- Place a larger stone, such as an amethyst cluster or a large piece of celestite, on the nightstand next to your child's bed.
- Massage your child's feet with a few drops of a stone elixir. Charoite is particularly effective in elixir form.
- Practice deep breathing, an age-appropriate meditation, or calming visualization while your child holds a stone in his or her hands.

Exercise for young children:

Create a magic box. Decorate it together with different symbols and pictures that evoke courage, light, and humor. If you want, you can put some sand in it with a couple of stones that always stay in the box.

Now encourage the child to imagine what his or her fear looks like. What color is it? What is the texture like? How big is it? Does it have a smell?

Together, put this "fear" into the magic box with a stone known to diminish fear and anxiety, such as amethyst. Ask the stone to help eliminate the fear, and imagine that it emits a bright light that penetrates and dissolves the fear. Reassure your child that it will continue to work as he or she sleeps. How have the color, size, and texture of the fear changed by morning?

❖ ❖ ❖

Bed-Wetting (Enuresis)

Ginette Tétreault

Enuresis, or bed-wetting as it is more commonly known, is the involuntary discharge of urine during sleep at an age after which a child has usually acquired bladder control. Studies have shown that some fifteen percent of children under age five suffer from this problem. The lives of affected families are significantly disrupted by the ensuing complications and repercussions. Parents may see their children's behavior change, their self-esteem and confidence shaken. They need to help their children deal with the psychological impact of this disorder by being understanding and minimizing its importance, as well as by making sure they know they are not alone in suffering from this condition.

Causes or Origins

Apart from a few rare cases of juvenile diabetes or an infection, enuresis is generally emotional and psychological in origin. Overly strict parental guidance, the birth of a new brother or sister, mistreatment, and emotional stress are the most frequently identified causes among children who suffer from nocturnal urinary incontinence. It has also been noted that such children are often extremely sensitive to their social and family environment. The emotions generally affect the bladder function, and the result can lead to an immature genitourinary system, which plays a role in enuresis.

Stones and Crystals to the Rescue

Various approaches are available to treat this problem. Thanks to the influence of the strong vibrations that the mineral realm exerts on the energetic meridians, it has proved to be a remarkable ally. What's more, it is a completely safe and effective option.

SUGGESTED STONES AND CRYSTAL

The main components of citrine, nephrite jade, prehnite, and sunstone are known for their capacity to impart healing energy to the kidneys and bladder, either for treating an infection or for incontinence.

Citrine

Yellow-gold in color, citrine is associated with the solar plexus located at the top of the sternum. It purifies the aura by filling the dark spaces with joy. It soothes family discord, encourages extroversion, and promotes self-expression. Its high iron, calcium, and manganese content fortifies the nerves, relieves bladder infections, and helps children overcome oppressive influences.

Nephrite Jade

Nephrite jade is a green stone that regulates the kidneys and urinary system. It balances the fluids within the body, helping to reduce incontinence.

Prehnite

Prehnite, a stone of unconditional love, seals the aura in a protective shield of divine energy, bringing tranquility, peace, and protection. Its vibration alleviates nightmares and deep-seated fears, healing the underlying ailment. Prehnite is recommended for hyperactive children as it helps eliminate the causes underlying this condition. Rich in calcium and aluminum, it treats kidney problems and urinary incontinence.

Sunstone

Sunstone is invaluable in supporting children during these distressing periods, dispelling stress, anxiety, and fear. Thanks to its affinity with the sun, it is a joyful, light-inspiring stone that instills good humor and boosts self-esteem. In addition, sunstone stimulates self-healing, regulates the autonomic nervous system, and ensures harmonious cooperation among the organs. Its healing properties purify the kidneys, bladder, and intestines.

How to Use These Stones

To ensure their child benefits from the full potential of these stones and crystals, parents should place them near the *hara* region, about one inch below the navel, which corresponds to the genitourinary system, for fifteen to twenty minutes. Because of these stones' affinity with this center, their energy can effectively irradiate to the bladder region.

Parents can also tape the stone to the child's lower abdomen before bedtime. If the child appears to be uncomfortable, stone elixirs, which have proved to be astonishingly effective, may be used. Simply place a few drops of the appropriate elixir in the palm of the hand and then rub the child's lower abdomen or back.

Achieving Energy Balance

SUGGESTED STONES AND CRYSTAL

Amazonite, Chrysoprase, Peridot

Green stones like amazonite, chrysoprase and peridot are also excellent choices because of their affinity with this energy center. Amazonite is particularly indicated if the incontinence derives from feelings of sadness or failure. Chrysoprase and peridot have a soothing effect, especially if the condition is due to a lack of attention or jealousy.

Rose Quartz

To promote the circulation of energy from bottom to top during lithotherapy treatment, the use of a stone or an elixir such as rose quartz at the level of the heart will transmit tenderness and calm as well as foster love and encouragement.

Consulting a qualified lithotherapist is an important step in finding the answers to the questions this condition raises. To ensure that this period of incontinence is as brief as possible, the insightful advice of such an expert will be appreciated by both parents and child.

◆ ◆ ◆

Urinary Tract Infections

Joani Gagnon

Although urinary tract infections (UTIs) mainly affect adults, children may also be afflicted. Some three percent of young girls and one percent of young boys suffer from a UTI before age eleven. It is extremely important to treat this infection in children since it can have lasting effects.

The Urinary Tract

The urinary tract, which is a crucial system in the body, is composed of the two kidneys, two ureters, the bladder, and the urethra. The kidneys filter out waste and water from blood, converting them into urine, which is conveyed to the bladder through two small tubes called ureters. The bladder temporarily collects and holds the urine. When the bladder muscle contracts, the urine is conveyed to the urethra, which produces urination.

Bacteria

Since the female ureter is shorter than the male's, bacteria from the skin, the genitals, and even the anus can easily infiltrate it. This anatomical difference means that girls are more at risk than boys for urinary tract infections. Bacteria can reach the bladder through the ureters and eventually reach the kidneys. Cystitis, a bladder infection, is the most common disorder, while urethritis, an infection of the ureter, and pyelonephritis, an infection of the kidneys, are rare among children, except for those born with renal weakness.

Cystitis

Cystitis is an inflammation of the bladder that may appear during the first year of life. Its symptoms include isolated fever, digestive disorders, lack of appetite, weight loss, and a generally altered condition. If you notice that your child has these symptoms, you should consult a specialist as soon as possible to ask for a urine analysis.

Children over age one who have a urinary infection exhibit the same symptoms as adults: an itchy irritation in the ureter, the need to pass urine more often, and a burning sensation after urination. Their urine may be cloudy and even contain blood. Bed-wetting is another sign that can help parents identify cystitis.

SUGGESTED STONES

Amber, Carnelian

As soon as the first symptoms appear, we suggest placing amber on the hara, which is located at a distance of two fingers below the navel, for children under five. For older children, placing a carnelian on the same area can also be effective. In both cases, leave the stone on for about ten minutes.

Malachite

Malachite is next on the list as it helps soothe the inflammation and purify the body at the deepest level. In addition, thanks to its green color, this stone promotes regeneration.

Urethritis and Pyelonephritis

If we don't pick up on the initial symptoms of our child's urinary infection, it can develop into urethritis, an inflammation of the ureters—the ducts through which the urine passes to the bladder—or even pyelonephritis, which is an inflammation of the kidneys. Happily, however, these two disorders are less common.

Among newborns, boys are usually more likely to have a weak urinary system, and these infections are often due to predisposing physical abnormalities. Among older children, the symptoms are the same as those of cystitis, accompanied by a high fever, shivering, increased sweating, and back and abdominal pain. If a child exhibits these symptoms, parents are strongly recommended to consult a physician and administer the prescribed medication.

SUGGESTED STONES AND ELIXIR

Aventurine, Emerald, Hematite, Prehnite

Stones can be a useful supplement to medication. Placing prehnite on the kidneys and applying a few drops of emerald elixir to the thymus help combat infection. If the child has a fever, hematite can lower body temperature and reduce stress and anxiety. In addition, placing aventurine on the solar plexus soothes and provides emotional support.

Prevention

To effectively prevent urinary infections, we need to teach our children to keep their genital organs clean. Girls should learn how to wipe themselves from front to back to prevent bacteria from the stools entering the ureter. They should also go to the toilet as soon as they feel the need to urinate and drink a lot of water, which will promote drainage and regularly cleanse the urinary tract. Using prehnite as a preventative remedy also helps cleanse the kidneys deep down.

Lastly, if the symptoms of any of these infections persist, parents are strongly advised to consult a specialist.

◆ ◆ ◆

Autism

Elianne Meier

Children with autism, one of the pervasive development disorders (PDD), are affected on numerous levels. In general, autism is characterized by verbal and nonverbal communication problems, difficulty in engaging in relationships with others, repetitive behavior, or limited interests that lead to rigid attitudes. Autistic children have trouble seeing the world as it is. Those working with autistics maintain that they are physically "here" even if they appear to be "elsewhere," although it is clear that they do not experience the same reality as most people.

The American child psychiatrist Leo Kanner was the first to identify autism as a pervasive development disorder in 1943. Given that it is a "recently discovered" neurological disorder, research to determine its causes is ongoing, and there are still numerous avenues to explore to improve this condition that will no doubt provide more information in the future. Although there is no cure for autism at present, education can play a major role in improving social and communication skills and in diminishing disturbing behavior.

Suggested Stones

Apatite

Apatite fosters communication. It encourages extrovert behavior, the awakening of consciousness, and a feeling of well-being in social settings. It enhances group communication and promotes the development of an accommodating attitude. It also dispels confusion, irritability, anger, frustration, and hyperactivity.

Blue Lace Agate

Blue lace agate calms the emotions, soothes tension, neutralizes anger, and facilitates communication in difficult situations. It introduces a positive attitude into the environment and promotes acceptance.

In ancient times, agates were considered lucky charms and stones of protection.

Charoite

Charoite is another beneficial stone for autistic children since it anchors them in day-to-day reality. It also mitigates feelings of alienation and frustration. A stone of transformation, it stimulates the latent potential in each of us. It helps overcome compulsions and obsessions while reducing stress and anxiety. Charoite opens the heart to the problems and suffering of others and invites us to accept people just as they are.

Sugilite

The properties of sugilite make it an invaluable aid for autistic children since it is recommended for those who don't feel "at home" on the earth. It anchors the soul in the reality of the present moment and helps autistics adapt to earthly vibrations. It enables us to open the door to our deepest thoughts and motivations. Sugilite also helps us believe in life and its manifold possibilities. A source of inspiration, it encourages us to change how we are and how we act. It banishes negativity, calms the nervous system, and reduces anxiety. Furthermore, sugilite reduces learning difficulties, diminishes emotional confusion, and banishes feelings of hostility.

Interestingly, sugilite and charoite are New Age stones that were discovered in the 1940s, as was autism, creating a mysterious synchronicity. This is just another example of how Nature offers us support, even for modern challenges.

SUGGESTED STONES
FOR THOSE WORKING OR LIVING WITH AN AUTISTIC CHILD

Lepidolite

There's no doubt that stress levels are higher for those who work or live with an autistic child, and they may sometimes feel exasperated and discouraged.

Lepidolite is one of the most powerful stones for relieving stress and eliminating negative thoughts. It makes it easier to accept the upheavals of life and allows the heart to be heard. A calming stone, lepidolite encourages us to confront challenging situations and helps us both see the appropriate course of action and discover hidden possibilities. It teaches us to listen to others with patience and compassion. While developing our inner strength and openness, it motivates us to focus on what is important in life and see its beauty, even in a negative environment. Lepidolite promotes in-depth emotional healing, helping us to accept the present. It is also recommended for children as it has many properties that can benefit them.

Autism can't be "cured" by stones, but some of its symptoms can be significantly reduced. The mineral kingdom can be a highly effective tool in encouraging autistic children to open up to the outside world and thus help them to thrive and grow.

PART 2: THE SCHOOL YEARS

Overcoming Anxiety
on the First Day of School

Jacqueline D. Sylvain

Our child's very first day of school is a milestone experience. Leaving their cozy nest, the family cocoon where they have learned everything they know and feel loved and safe, some children are extremely anxious and emotional as they enter this new phase of their lives.

As parents, we are obviously affected by our children's feelings and try to put ourselves in their place. Added to all this emotion is our own experience, be it good or bad, of our first day at school. We wonder how we can calmly prepare and reassure our children as they set off on this great adventure.

Suggested Stones and Crystals for Parents

As the first day of school approaches, we have to be able to live up to our responsibilities as parents and communicate a sense of calm and self-confidence to our children. We should be able to coach them in their new activity and provide them with loving support, while giving them the space they need to experience this new step in their own way.

Amethyst, Charoite

Amethyst and charoite help us deal with the worries, stresses, and fears we face. Amethyst, offering powerful protection, calms anxieties and fears, and charoite takes us beyond these fears, imparting determination, spontaneity, and efficiency.

Labradorite

As parents, we have to stay centered and in control. Labradorite is an ideal choice in this situation. It protects us from negative energies of all kinds, helping us conserve our energy and "keep cool." It also enables us to eliminate our fears and insecurities and boosts our self-confidence.

Larimar

Our emotional bodies may have to work overtime during this period. The stabilizing effect of larimar encourages us to stay calm during periods of stress and change. It is also especially useful for alleviating feelings of guilt and allows us to let ourselves be gently carried along by the current of life.

Parents' kit: amethyst, charoite, labradorite, larimar

SUGGESTED STONES AND CRYSTALS FOR CHILDREN

When our children anxiously clutch us by the hand and beg us not to leave them in this unfamiliar place where every face is strange to them, the memories of our own first day at school come flooding back. How can we help them and make them feel secure? How can we make them understand that we still love them and aren't abandoning them?

Amethyst

Amethyst should definitely be on the agenda. It eliminates feelings of anxiety in the face of the unknown and inspires the courage to cope with the trials and tribulations of life.

Aquamarine

Aquamarine is a useful stone in this situation as it confers the courage to confront this new world and fosters self-expression.

Citrine

Citrine is a warming stone that brings joy to everyone who wears it. It also promotes the development of positive attitudes.

Rose Quartz

Rose quartz is most important when love appears to be in doubt and we suffer from a feeling of abandonment. It also reassures and calms, emitting loving vibrations and opening hearts to be receptive. Rose quartz facilitates the acceptance of change, so necessary at this time, and is particularly suited to children because of its gentle, soothing action.

Children's kit: amethyst, aquamarine, citrine, rose quartz

SUGGESTED STONES FOR TEENS

When back-to-school season rolls around, many teenagers are also apprehensive about going back to class, especially when they first enter high school, where everything is new. They have to get used to a new environment, make new friends, get to know their teachers, learn to comply with new regulations, and even overcome their fear of the school bus. What can we as parents offer our teenagers to help them get through the rest of the year? Image is everything for teens. Their "look" and how they dress have a positive or negative impact on their self-confidence and thus influence whether they are accepted or rejected by their peers.

Lapis Lazuli

Teens who want to be universally loved and accepted at any cost must learn to distance themselves from this vital need and rediscover the inner strength and dignity that will help reduce their dependence on others. Lapis lazuli will inspire them to take charge of their lives and be independent. Known as a powerful protective stone, it encourages us to take control of our lives. It also promotes honesty, compassion, and integrity.

Ruby

Ruby, another powerful stone, instills respect for the person wearing it. It has a positive impact on those who lack willpower and tend to underestimate themselves. It imparts the courage to exceed limits while keeping an open heart. However, it is not recommended for strong or arrogant personalities, which are often common at this age.

Sodalite

In contrast, sodalite is highly recommended for strong or arrogant personalities. A stone of humility, it fosters harmony and solidarity and supports group work. It brings emotional balance and calms anxiety. It also promotes expression, intuition, and open-mindedness, and it encourages us to let go of rigid, outmoded mental conditioning. In addition, sodalite increases self-respect, self-acceptance, and self-confidence.

Sugilite

Sugilite is a wise choice for young people who feel different, out of step, and have trouble being part of the crowd. It alleviates fears and paranoia, and eliminates hostility while promoting forgiveness. It can help resolve conflicts, contributes to group work, and encourages loving communication.

Teens' kit: lapis lazuli, ruby, sodalite, sugilite

Stones and crystals are simple and safe to use. They bring us support, strength, and comfort. Experience has shown that these precious allies work to assist us at all times in our lives. Their presence is invaluable in helping us reestablish and maintain our balance so that everyone benefits.

GROUP INFLUENCE

PHILIPPE SOREAU

How to Preserve Your Identity, Survive, and Avoid Becoming a Victim

Belonging to a group and feeling connected to others who share the same interests, attitudes, and circumstances is important for our development and that of our children. As children grow up, they increasingly feel the need to belong to a group and be recognized. Some children are easily swayed by a group. The influence of their peers, either positive or negative, is of major importance in their lives; in fact, the opinions of the other group members often carry more weight than those of their parents.

How can we help our children have more control over their lives and clearly understand the outside influences of the groups to which they belong? Parents can provide support by showing their children that they love them, establishing clear boundaries, encouraging them, and discussing with them what's happening in their lives each and every day. Some stones can also be very beneficial in helping children develop qualities that will enable them to more effectively integrate their experiences, learn from them, and grow.

Identity and Independence

It is important for children to attain a certain level of independence and establish a sense of personal identity to help them interact with the other members of their group.

SUGGESTED STONES

Lepidolite

Lepidolite promotes autonomy. With its support, children can more easily set their objectives and attain them without help from others. It also protects against outside influences and helps us preserve our identity in groups. In addition, it helps us overcome all kinds of emotional and mental addictions.

Sugilite

Sugilite, a violet-pink stone, helps us live meaningfully and according to our own truth. Its energy encourages us to concentrate on what is essential and avoid distractions. It alleviates sorrow, grief, and fear. It is also useful in resolving group conflicts, eliminating hostility, and inspiring solutions that meet everyone's needs fairly. It is highly appropriate for group work, as it encourages forgiveness and loving communication.

Self-confidence and Self-assurance

Children who are self-confident and self-assured find it easier to interact in a group. The following stones promote the development of these qualities.

SUGGESTED STONES

Agate

A stone of protection, agate can help children banish negativity. It inspires strength and courage, fostering tranquility, balance, and harmony. It reinforces powers of concentration and brings clarity of thought, thus helping them discern the truth and accept reality.

Aventurine

Aventurine strengthens leadership qualities and promotes empathy and compassion. It relieves stress and confers the courage to live our truth from the heart. It also has a balancing effect, stabilizing a child's mood and enhancing creativity.

Fluorite

Fluorite reinforces concentration and favors swift comprehension and quick thinking. It also helps dispel confusion and stimulates inspiration, creativity, and learning. A highly protective stone, particularly on the psychological level, fluorite reduces negative energy and stress. It is indispensable for bringing new awareness to any situation, enhancing the understanding of others, and thus improving a child's cooperation within the group.

Tiger Eye

Another protective stone, tiger eye is essential in this case since it reduces the impact of oppressive and threatening situations children can experience. It also helps dissolve fear and anxiety and stabilizes emotions while conferring courage, strength, and determination.

Humor and Joy: Key Assets for Maintaining Balance and Harmony

SUGGESTED STONE AND CRYSTALS

Citrine

Citrine imparts joy and good humor. It alleviates tension and discord, helping children overcome oppressive influences and stimulating self-expression. It hones creativity, encourages new ideas, and promotes the resolution of typical group problems. In addition, it fosters the desire for change and self-realization.

Labradorite

Another powerful stone of protection, labradorite protects children by forming a barrier against negative outside influences. It acts as a bridge to universal energies, elevates consciousness, and encourages more profound feelings. It enhances our ability to please others, helps develop friendships, and nourishes those who feel alone.

Herkimer Quartz

Herkimer quartz can also be very useful. Like citrine, this quartz is known for bringing joy and harmony and encouraging creativity. It helps eliminate negativity and facilitates liberation and transformation.

Self-expression and Communication

The ability to communicate and express ourselves is a major asset in learning to effectively interact within a group.

SUGGESTED STONES

Aquamarine, Lapis Lazuli, Sodalite

Aquamarine, lapis lazuli, and sodalite are all stones that facilitate communication and the verbal expression of feelings and emotions. They inspire the search for truth; bring calm, lightheartedness, and serenity; and harmoniously balance the emotions. These stones stimulate mental development, opening us up to the bigger picture and thus inspiring tolerance and increasing presence of mind. As such, they are highly beneficial to group work, since they stimulate confidence, communication, discussion, negotiation, camaraderie, harmony, and solidarity while helping to translate ideas into reality and to attain group objectives.

Ruby

Ruby is also an excellent stone for imparting strength and enabling us to take our rightful place within a group without losing our integrity. This precious stone inspires self-respect and the respect of others.

There are numerous stones available to sustain children in group interaction. Depending on what they are looking for, they can choose the stones and crystals that most effectively meet their needs. These companions will confer qualities that will enable them to make the most of their group experiences and ensure that both they and the group flourish.

How to Triumph
Over Bullying at School

François Nicol

The arrival of fall coincides with the end of the summer vacation for thousands of young people who will be returning to school and hitting the books. Many look forward to going back to school because they'll be reunited with their friends. Others see it as a time when they'll have to face new and unknown challenges that create considerable stress. Unfortunately for some children, going back to school means returning to a taxing and intimidating nightmare. As victims of bullying, these children live in a world where insults, physical aggression, fear, and rejection are day-to-day realities. Most of the time, they feel powerless and have no idea how to cope. How can we as parents help our children deal with these stressful situations?

Build Their Self-Confidence

Self-Confidence is the key that will enable children to triumph over these distressing experiences. It is self-confidence that will give them the strength they need to stop the harassment that is making their life at school so miserable.

Garnet, Ruby, Tiger Eye

Garnet, ruby, and tiger eye, which are known as warrior stones, are ideal for building this confidence, acquiring strength, and learning to express the desire for respect. These stones will encourage children to take control in problematic situations, thereby gradually reducing their identification with the role of victim. Lithotherapists recommend choosing a stone that can be worn as jewelry or slipped into a pocket so that its benefits can be felt all day long. Three drops of a stone elixir two or three times a day can also be beneficial. However, if the child becomes angry and impatient, the dose should be decreased.

Assertiveness

For children to assert themselves, they must have acquired some measure of self-confidence. Without being rude, they have to make it clear to their aggressor that they don't appreciate his little game. Being assertive is both effective and liberating. By expressing themselves, bullied children articulate the feelings they've been repressing for too long. They will then feel lighter and more comfortable within themselves.

SUGGESTED STONES

Apatite, Aquamarine

Apatite and aquamarine act directly on the throat, the site of self-expression, and therefore effectively contribute to the child regaining control. These stones also promote clear and creative communication while granting the courage needed to resolve difficult situations.

Alleviating Anger

Bullying destabilizes a child's psychological balance. Feeling threatened, either for short or long periods of time, can often result in various behavior problems for the victim. Aggression, rejection, lack of respect, and the constant suppression of emotions frequently lead to negative feelings. The anger sparked by this oppression often translates into hatred of the aggressor.

SUGGESTED STONES AND CRYSTALS

Ametrine, Citrine

Ametrine and citrine inspire the joy and lightheartedness that are necessary for feeling good at school. They stimulate concentration and alleviate anxiety, and they encourage the formation of relationships and playing a more active role in group situations. They also help put things in perspective, enhancing our comprehension of the situation and our ability to find solutions.

Aventurine, Kunzite, Rose Quartz

Aventurine, kunzite, and rose quartz soothe intense emotions and create a feeling of calm that is conducive to relaxation and understanding. These stones encourage forgiveness and teach us how to love ourselves again. They help children feel less isolated and more appreciated and loved at school. When they feel more confident, they are less afraid of connecting with others.

Parental Support

It is vital for parents to understand the pain their children are experiencing and to help and support them as they strive to be more assertive. Parents play an indispensable role in ensuring that their children recover their balance and personal well-being.

SUGGESTED STONE

Dioptase

Dioptase will help parents remain aware of their children's needs and develop the empathy to truly understand their suffering. As a result, their children will feel supported and confident. We must not forget that parents are the pillars that sustain children at all times.

Easy to use and effective, stones and crystals are always prepared to help anyone with problems. There is no question that they can be exceptional caretakers for children being bullied at school. Parents must remain vigilant and avoid trivializing the situation; they can then call on the wisdom of stones and crystals for help. Thus it will eventually be possible for children to find their own particular niche at school and, above all, be happy to be heading back to the classroom.

◆ ◆ ◆

Developing a Sense of Organization

Annie Dufresne

Today's world moves so fast that we often feel out of breath, as if we have been running a marathon but never reach the finish line. Our children's days go by at a hectic pace, and many of them have trouble organizing their thoughts, let alone their environment. How can we help our children develop a sense of organization in all aspects of their lives? At school and at home, a sense of order and balance can improve their inner lives as well as their relationships. Fortunately, several stones that radiate these qualities are available to us.

Suggested Stones

Fluorite

Fluorite is an indispensable aid in helping us organize the information we receive. It enhances concentration and stimulates rational thought, encouraging children to tackle tasks one by one, in logical fashion. Fluorite imparts the confidence to meet day-to-day challenges. Why not add a few drops of fluorite elixir to your child's water bottle? It could be invaluable for dealing with high stress levels at school and improving thought processes and mental concentration.

Jasper

Jasper, which comes in a variety of shapes and colors, is particularly appreciated by children. It promotes organizational skills and encourages children to clearly express their ideas and put them into action. Its contribution will help them feel grounded and increase order and routine in their daily lives.

Petrified Wood

Petrified wood seems to be made to order for children who lack focus. As varied in beauty as the trees in a lush forest, it uplifts the spirit and helps untangle the maze of life. This anchoring stone inspires us to live more simply and keep our feet firmly planted on the ground. It brings a sense of calm to children and wraps them in a feeling of gentle comfort. It can also stimulate their memory and bring daydreamers back to earth. What child could resist such a generous gift?

Pyrite

Pyrite attracts children of all ages by its brilliant gold color. This inviting, generous stone encourages organization, structure, and order. Like the warmth of the sun, it boosts our energy, joy, and sense of well-being each and every day, making it an ideal stone for children who often seem to be on another planet. Pyrite structures our mental faculties and helps us develop a beneficial sense of order. By balancing different parts of the brain, it can infuse logic into the whirlwind of creativity inside each child. However, children are not advised to wear the stone; simply placing it next to their desks will do the trick.

Sodalite

Rich blue in color, sodalite also stimulates logical thought since it holds the secrets of the perfect order of the universe deep within itself. It may well inspire some children to take their first steps toward better organizing their time, feelings, and ideas. Structuring both the brain and thought, sodalite fosters the practical abilities so many children need. Simply wearing a sodalite bracelet can bring about these changes.

A sense of organization is a skill that can be taught and learned. With gentle guidance and patience, children can gradually learn to organize their inner and outer worlds in a healthy way. May each child enjoy this extraordinary adventure!

◆ ◆ ◆

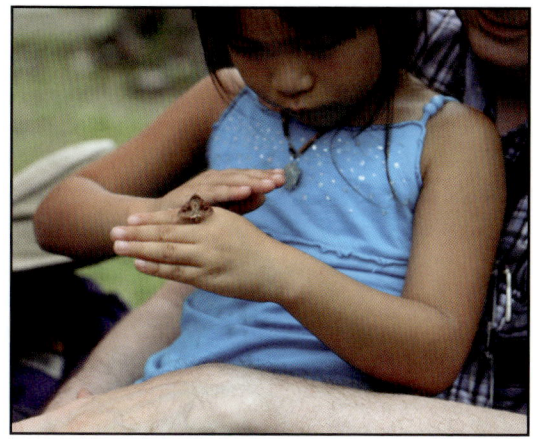 

CONCENTRATION

Laëtitia Betton

Although we're not all born with the same ability to concentrate, this faculty can be developed to varying degrees, according to our individual interests. Children are known for being able to quickly tune out any subject that bores them, either by getting lost in thought or escaping into play. Stones can step in to focus their attention and lend assistance when children have to study or work on something they find less than appealing. They encourage children to live in the present by promoting concentration and helping with the assimilation of information. The better children concentrate, the more they remember. Here are a few stones and crystals that can support our budding geniuses.

Stimulating the Brain and Memory

SUGGESTED STONES

Fluorite

Fluorite is very useful when action is required or when a situation calls for impartial and objective judgment. It promotes analytical skills and improves concentration. Fluorite balances our mental faculties and clarifies and stabilizes our emotions. It reduces stress and strengthens the memory as well. This stone stimulates the desire to learn, promotes understanding, and inspires quick and logical thinking. Fluorite develops decision-making skills and helps make children more responsible.

Pyrite

Pyrite builds self-confidence. Thanks to its iron content, it brings blood to the brain, thereby stimulating mental activity and concentration. It increases energy and helps overcome the fatigue children often feel when they have to study. This stone stimulates the memory and intellectual faculties, enabling children to draw on their unexplored ability and potential. Psychologically, pyrite soothes anxiety and the frustration that young people can feel when they don't understand a concept, a formula, or a difficult lesson.

Sodalite

Sodalite is a must for hypersensitive children. It helps them gain better control over their emotions and gently grounds those whose heads are often in the clouds. This stone is tied to intellectual endeavor, being synonymous with stability, logic, rationality, and self-control. It enhances the sense of logic and is suitable for children who are naturally attracted to the sciences as well as those who dislike mathematics. In addition, it develops concentration and brings the inner peace that will help the personality flourish. It is essential for children since it helps them express who they are. Sodalite promotes self-esteem, inspires confidence, and encourages solidarity. It is also ideal for supporting group work or school projects.

Fluorite, pyrite and sodalite are all stones belonging to the cubic alliance,[1] which means that the molecules of these stones bind together to form a cubic structure. This shape promotes order, structure, and logic in children. The cubic alliance is closely tied to the development of concentration and all aspects of school and learning.

---

[1] The study of the geometric alliances is discussed in The Group of 5, *The Eight Crystal Alliances: The Influence of Stones on the Personality* (Montreal: Paume de Saint-Germain Publishing and Berkeley, CA: North Atlantic Books, 2010).

Protection from Outside Influences

SUGGESTED STONE

Lepidolite

Lepidolite is ideal for helping children to concentrate on the matter at hand and to avoid distractions. This stone protects children from outside influences and helps maintain their powers of concentration. Stimulating the intellect, it encourages objective and wise decision making. Its lithium content promotes balance and calms the mind. Lepidolite is also effective for bipolar disorder and chronic anxiety.

Learning to Concentrate

SUGGESTED STONE AND CRYSTAL

Carnelian

Carnelian is highly appropriate for lethargic children. Its energizing action stimulates the brain and promotes concentration. It eliminates confusion and facilitates problem solving. Carnelian also inspires the courage to believe in ourselves and heightens creativity.

Citrine

Citrine is remarkable for increasing concentration. Its luminosity helps streamline children's thoughts and enables them to focus on the subjects discussed in class. Since it also promotes digestion, primarily by acting on the liver, it clears the mind. A properly functioning liver leads to improved vision and reduces mental fatigue. Furthermore, citrine is a mood elevator and helps children overcome worry and stress.

It is comforting to know that stones can serve children on many different levels, particularly when it comes to concentration. They gently impart their power, giving children a sense of well-being and stimulating their intellect.

◆ ◆ ◆

MANAGING AND RESOLVING CONFLICTS

ANNIE DUFRESNE

Many of us suffer from unresolved conflicts, and children are no exception. From early childhood, they have to develop healthy strategies that help them address these conflicts fairly and with a sense of balance. Discord is part of the human experience, which is why it is particularly important to ensure our children are equipped to take their places in an often chaotic day-to-day reality. Luckily, a number of stones can provide calm, comfort, and support during this learning experience.

SUGGESTED STONES

Aventurine

Aventurine encourages us to be tolerant toward others and to understand them better. It alleviates the emotional stress experienced by some children when faced with conflict and fulfills those who don't feel loved. A gentle green stone, it lovingly sustains all who wear it.

Epidote

Olive green in color, epidote promotes positive interaction with others. It diminishes tendencies to criticize ourselves and others, smoothing our relations. Just as certain foods sustain our physical abilities, epidote can accomplish wonders by nurturing children's relationship skills. A wise companion, this stone guides children in choosing their friends, helps them maintain healthy self-esteem, and inspires long-lasting joy in living.

Jasper

Known in the Middle Ages as the stone of warriors, jasper protects the fearless warrior within. It supports children in times of conflict, secretly imparting courage and determination. Like invincible yet invisible armor, this stone keeps our little troopers safe.

Labradorite

Labradorite is a source of light and encouragement when conflicts persist and affect a child's well-being. It imparts strength and perseverance and forms a barrier against unwanted energies. Like a shield, labradorite protects children by preventing them from absorbing the negative energies of others.

Rhodonite

Rhodonite can support children who feel bitter after a quarrel. Thanks to its pink color, this stone emanates the powerful energy of love that brings emotional balance and gently dissolves resentment. It encourages brotherhood and forgiveness, inspiring important insight during conflicts. When children lose hope about resolving conflicts in their day-to-day lives, rhodonite will guide them toward the solution to seemingly insoluble problems.

Sugilite

Sugilite alleviates sorrow and fear by inspiring self-forgiveness. It calms discord and encourages us to find solutions that fairly answer everyone's needs. This stone promotes forgiveness, helps banish hostility, and fosters group harmony. Sugilite is a powerful ally in restoring harmonious relations within a family, a group, or in the classroom.

Whatever conflicts our children experience, they can become a valuable source of learning for younger and older children alike. Once they have understood these conflicts, they will then have the opportunity to improve and open their hearts to the suffering of others. We must learn to trust our children's natural innocence, for that's where the real potential lies to create healthy long-lasting relationships.

◆ ◆ ◆

Preparing for Exams

Johanne Marier

Children are often extremely anxious when they prepare for their exams. They feel they have to pass and are worried about not being able to meet their parents' expectations and disappointing them. The first tip for parents is not to pile on any additional pressure. Of course, we have to help our children and teach them the importance of making an effort. However, we shouldn't forget that some of them will never meet the excessively high standards their parents set, which can make them feel discouraged and depressed. Everyone works at their own pace. Children shouldn't identify with their grades; they are indicative of only one small aspect of their accomplishments. People who are successful in life aren't necessarily the ones who had the best grades, but rather the ones who learned that effort and daily work are the keys to success. So how can we help steer our children calmly through the stressful examination period?

Aligning the Bodies

It's important to help our children with both their inner and their academic preparation. For them to achieve mental clarity, their bodies must be correctly aligned in order to allow the proper flow of energy.

SUGGESTED STONES

Kyanite

Kyanite, which instantly aligns the chakras and balances the body's energies, is definitely on the agenda at exam time. It strengthens the functioning of the cerebellum and motor nerves, thus improving mobility and agility.

Sodalite

Sodalite is useful for calming children before they begin to study for an exam. A stone that stimulates logical and rational thought, it also stabilizes the emotions, counteracting rising anxiety and apprehension. Worn in a bead necklace under a T-shirt, its effectiveness increases tenfold; it can also be highly effective when slipped into a pocket or placed on a desk or dresser in the bedroom.

Overcoming Anxiety

Once the inner preparation has begun, there are other important factors that parents should focus on: their children's anxiety level, self-confidence, and intellect. Parents can help—and so can stones.

SUGGESTED STONES

Blue stones

Blue stones provide strong support to anxious children through their calming qualities. No matter which blue stone a child may choose, it is sure to help, since children are intuitive and know how to choose wisely.

Rhodochrosite, Rhodonite

Children who are anxious the night before an exam often have trouble sleeping and may even feel physical discomfort. Parents can alleviate these problems with stones like rhodochrosite and rhodonite, which will help their child manage his or her emotions and pre-exam stress.

Acquiring Self-Confidence

A healthy dose of self-confidence can be very useful in preparing the ground for the learning process. In fact, self-confidence often eliminates pre-exam jitters. Which stones can reinforce this assurance? A few effective possibilities are discussed below.

SUGGESTED STONES

Chrysoprase

Chrysoprase is a remarkable stone that can calm the nerves and bring mental clarity.

Orange Calcite

Orange calcite is a highly positive, luminous, and warm stone that increases confidence. Children can hold it in their hands until they feel its benefits, or slip it under their pillow the night before the exam.

Smithsonite

Smithsonite is a stone that offers beneficial support for prepubescent girls and teens. It helps them acquire inner certainty when confronted with the apprehension and doubt that often surface during this period.

Stimulating the Intellect

Does your child need help boosting memory, stimulating the brain, and maximizing his or her intellectual faculties, all important assets when preparing for an exam?

SUGGESTED STONES AND CRYSTAL

Citrine

Placed in a desk or study room, citrine promotes the review of knowledge and learning. It also reinforces concentration. An extremely positive stone, it stimulates the mind, dispels fatigue, and imparts joy, which makes it ideal for students.

Fluorite

Fluorite is indispensable during the examination period since it helps us acquire rational thought processes and facilitates concentration.

Pyrite

Pyrite tops the list for its ability to stimulate the intellect. It is excellent in helping improve the memory. It also imparts organization, structure, and order, three key qualities at exam time.

If you can't make up your mind which of these stones to choose, why not let your child make the decision? And if he or she asks you to choose, listen to your intuition. The stones will be visibly effective, and your child will feel better in a very short time. Once you have assembled a kit of sustaining stones, you'll all be able to say good-bye to pre-exam stress.

♦ ♦ ♦

Oral Presentations

Laëtitia Betton

For most children, and even for adults, speaking in public can be stressful and unnerving, especially if the experience brings to mind painful traumatic events from the past. Sweaty palms, a dry throat, and butterflies in the stomach are only some of the symptoms of stress that oral presentations can cause. Developing self-confidence is indispensable to alleviating this distress and putting a positive spin on these emotions. A number of stones have the qualities needed to comfort and sustain children and teens in this situation. It is simply a matter of selecting the stone that best suits each child's personality.

Suggested Stone for Rebellious or Angry Youths

Blue Lace Agate
Blue agate, also known as blue lace agate, has a very gentle energy. It brings soothing calm to aggressive children and rebellious teens. It neutralizes anger, alleviates fever, and counteracts the fear of being judged and rejected, which is often present in a school environment. Blue lace agate encourages understanding and contentment. It also fosters diplomatic communication, encourages expression, and is particularly helpful in reducing stuttering.

Suggested Stone for Artists

Aquamarine

Aquamarine promotes oral and artistic expression. It is an excellent stone for opening communication with others and supports expression in all its forms, from public speaking to singing, writing, poetry, and philosophy. Highly suitable for artistic temperaments, it imparts inspiration and efficiency, banishing anxiety and gloomy thoughts.

Suggested Stone for Shy People

Blue Calcite

Blue calcite fosters clear communication and is appropriate for introverted or shy people. It encourages them to overcome and master their fears. By alleviating their worries, it makes it easier for them to reach out to others. Blue calcite also encourages positive thinking and increases personal energy.

Suggested Stone for Erasing Bad Memories

Lapis Lazuli

Lapis lazuli offers support when children or teens seek solutions and direction within themselves as part of the creative process. It teaches the power of words, counteracting the distress caused by past lack of expression. Lapis lazuli will help children express their opinions, feelings, and emotions. It inspires integrity, making them feel whole and able to reveal their true selves to others.

Suggested Stone for Complainers and "Victims"

Turquoise

Turquoise bestows confidence and the firmness of mind necessary to present clear, concise arguments. It sustains expression and encourages openness to the ideas of others. It also protects against disturbing dreams and nightmares, which is particularly helpful the day before a new school term begins. As well, turquoise prevents panic attacks and is appropriate for children and teens who tend to complain or have a martyr complex. It provides strong protection against all negative behavior, including self-destructive behavior.

The way to use these stones is very simple. For example, they can be worn as a pendant directly on the skin or taken as an elixir in the form of three drops under the tongue before meals. All these stones provide excellent support for oral presentations. They stimulate expression and can calm the nerves of young children, teenagers, and adults alike.

◆ ◆ ◆

Children Suffering from Rejection

Bertrand Corbeil

No one wants to be rejected, and no one can claim that they're not deeply affected when it happens to them. Although rejection is nothing new, when our children are on the receiving end, we feel as miserable as they do. Two types of children trigger rejection: those whose personalities, offensive behavior, and antisocial actions generate hostility, and those who have done nothing whatsoever to deserve it. Children in the latter group may be small for their age, obese, shy, anxious, or immature. But what matters is that they are different, and that difference exposes them to ridicule.

Children who are rejected or bullied at school are often stressed and lack motivation. They may sometimes act out their aggression, have trouble sleeping, lose their appetite, cry for no reason, or even become depressed. It is up to their parents to be attentive to these behavior changes and to try to find out what's wrong. Asking them questions about their friends, about what happens on the playground or in the lunch room, for instance, is usually the only way to determine what's going on. These children often need to be comforted and feel that they're being listened to, supported, and loved.

SUGGESTED STONES AND CRYSTAL

Rhodochrosite, rhodonite, and rose quartz are three allies from the mineral realm that can support and sustain children who suffer from rejection. Pink, a color related to the healing ray, has the power to raise vibrations and provide a wealth of benefits.

Rhodochrosite

Rhodochrosite has a calming effect, deepens breathing, and helps us control emotional turmoil by reducing emotional stress. It encourages us to express our feelings and face up to our fears. This stone helps children become aware of their own worth and gently brings repressed emotional wounds to the surface, enabling them to be recognized and then dismissed.

Rhodonite

Rhodonite imparts emotional balance and dissolves the resentment, anger, and pain caused by abuse. It restores peace of mind in frustrating and stressful situations. Rhodonite also encourages reconciliation, even after abuse.

Rose Quartz

Rose quartz provides reassurance and effectively alleviates emotional pain since it strengthens self-love. It gently absorbs negative energy, replacing it with loving vibrations. It teaches us how to love ourselves, which is crucial when we feel unworthy. Rose quartz envelops the heart in warmth and well-being. Relieving personal and environmental stress, it helps balance emotional energies and dispel anger and misunderstanding.

Black Stones

Apache tear obsidian and black obsidian are two highly protective stones. They can, however, expose the darker side of our nature, where jealousy, anger, and possessiveness lie, compelling us to recognize these traits and work to overcome them. If parents see their children becoming angrier and more frustrated, placing these stones next to them will help calm these emotions.

Apache Tear Obsidian

Apache tear obsidian, which has a more soothing effect than black obsidian, provides gentle protection. Although it has the same properties as black obsidian, it is more appropriate for children because of its gentler nature.

Black Obsidian

Black obsidian creates an invisible protective shield around hypersensitive people. It helps dispel their fears, emotional shocks, traumas, and feelings of victimization.

Green Stone

Moldavite

Moldavite naturally attracts children, perhaps because of its "extraterrestrial" origins. It reestablishes harmony and guides us in our work, relationships, and lifestyle. It is a very useful stone for sensitive people who find it hard to accept their earthly incarnation and are unable to adjust to suffering and deep emotions. However, it should not be worn continually, as it can leave you feeling "spaced out." To counter this effect and promote grounding in the physical world, moldavite should be used with hematite or smoky quartz.

Rejection is undoubtedly one of life's most difficult experiences. But the mineral world offers help to children and adults alike in this situation. No matter what the problem, it provides support and teaches children to smile again, bringing joy to the hearts of their parents.

◆ ◆ ◆

THE ART OF SELF-ASSERTION

Eliane A. Panneton

Children need to know how to assert themselves. Through self-expression they can convey to the world around them who they really are. Being able to express themselves allows them to communicate their needs and feelings, such as sorrow, fear, or joy. This interaction gradually teaches them validation and self-knowledge, leading them to ultimately come into their own. Self-assurance is critical to communicating with confidence and at times with courage. The ability to express ourselves is also based on love and self-respect, which develop at a very early age. Asserting ourselves means expressing who we are—in other words, simply being.

SUGGESTED STONES AND CRYSTALS

Amazonite

Amazonite can help children who repress their thoughts and emotions either because they fear confrontation or fear being judged by others. It fosters their self-expression while helping them open up to others' points of view. Amazonite encourages children to take themselves in hand and tap into their inner strength.

Aquamarine

Aquamarine is a powerful stone for clearing and activating the throat chakra. It is excellent for children who have disassociated from their emotional body after a trauma. Like blue lace agate, it enables them to express their emotions clearly and accurately. Due to its soothing effect, it is very beneficial for reducing the stress of bashful children or those who have trouble expressing themselves. Aquamarine's main virtue is inspiring courage.

Blue Calcite

Blue calcite instills stability and self-confidence. It promotes clear communication and helps reduce stress.

Blue Lace Agate

Blue lace agate encourages shy children to express their needs and emotions and to be heard. It can break down blockages, such as the fear of being judged or a lack of confidence. It acts slowly and gently, creating a feeling of calm and security that fosters self-acceptance, which in turn gradually boosts self-confidence. Blue lace agate also promotes deep peace.

Blue Topaz

Blue topaz enables children to achieve self-realization through the discovery of their inner resources. It calms the emotional body and enhances children's ability to express their ideas, innermost feelings, and needs.

Citrine

Citrine gives children self-assurance, courage, and self-confidence, helping them to value themselves more and express their individuality. It also generates joy and light, alleviating sadness and stimulating creativity.

Kunzite

At times children's refusal or inability to express and assert themselves stems from an emotional trauma that causes them to turn inward for protection. Kunzite helps heal the emotional body and encourages the constructive expression of suppressed feelings. It opens the heart to the love of self and of others, fostering the development of self-respect and an enthusiastic, joyful approach to life.

Kyanite

Kyanite promotes creative personal expression and communication. It encourages children to affirm their own truth and to overcome fears and crippling barriers.

Lapis Lazuli

Lapis lazuli helps children know who they are, express their emotions, and discern and discover truth. By fostering communication, it invites them to forge new relationships.

Malachite

Malachite reduces shyness, releases inhibitions, and encourages the expression of emotions.

Pink Calcite

Pink calcite helps children recognize their own value and eliminates fear and sadness. Placed in a child's bedroom, it emits a soothing energy and promotes a loving atmosphere.

Rose Quartz

Rose quartz is invaluable in soothing emotional pain. It strengthens self-esteem and has a calming, reassuring effect.

Stones and crystals can help children assert themselves by restoring their self-confidence and breaking down barriers that block communication. They smooth a child's progress along the path toward self-knowledge, self-acceptance, and self-expression.

Dyslexia

Annie Dufresne

Winston Churchill, Ludwig van Beethoven, Benjamin Franklin, Steven Spielberg, Whoopi Goldberg, Isaac Newton, and Agatha Christie all had things in common: they marked history with their passion for life—and they all had dyslexia. Luckily, dealing with this condition in a positive and constructive way can help dyslexics give the best of themselves to a greater cause. Although in the final analysis the power lies within each child to make his or her dreams come true despite this disorder, parents nonetheless jump in to search for effective solutions when confronted with the very real problems a dyslexic child has to overcome. Some stones, which may be quite ordinary in appearance, have the unique gift of calming mental confusion and helping balance different parts of the brain. Let's discover some of them.

Suggested Stones

Apatite

Dyslexic children sometimes seem to be drowning in confusion. Whether at home or at school, they must continuously battle against their inner turmoil to get their work done. Apatite will motivate them and help them tap into their energy reserves, diminishing their lethargy and fatigue. They will then be able to give free rein to their intellect and creativity and express the unique facets of their personalities.

Carnelian

Carnelian, an excellent stone for stimulating concentration and imparting problem-solving skills, can be a reliable companion for schoolwork, in the classroom, at home, or anywhere else. What's more, it can be particularly helpful to children with learning difficulties since it stimulates the brain and sharpens the intellect. This unique stone is highly recommended for dyslexic children lost in a fog of letters, syllables, or words.

Fluorite

Fluorite is a valuable aid for dyslexic children because it promotes rational thought and mental clarity. In the classroom or anywhere they do their homework, a hint of fluorite essence can reinforce concentration. Put five to seven drops of this elixir in a vaporizer containing about 2¼ cups of water and diffuse it into your child's living environment to ensure that fluorite's strength will be used at critical moments in the learning process. It is also a powerful support for your child's study area.

Kyanite

Because children with dyslexia are also often multitalented, they can be creative and spontaneous. However, they have to overcome a host of obstacles. For example, they may frequently compare themselves to others who seem to have an easier life, or they may suffer from occasional bouts of low self-esteem. Kyanite, which is a soothing blue color, cuts through these issues and promotes creative expression. It dispels confusion, encouraging logical and rational thought. It reinforces the memory and confidence of dyslexic children, equipping them with the additional tools they need to achieve academic success.

Petrified Wood

One of nature's wonders, petrified wood can serve as a beacon for the agitated neurotransmitters that are central to memory. It helps absent-minded children to focus and favors their intellectual growth. Closely linked to the earth, petrified wood grounds and anchors its wearers, encouraging them to feel they're in the right place at the right time.

Sugilite

Sugilite eases psychological tension and helps overcome learning difficulties. It soothes the nerves and brain in general and relieves the melancholy that this disorder can trigger.

Tourmaline

Lastly, tourmaline, which is known to be effective in treating dyslexia, balances the highly active cerebral hemispheres of children at the learning stage in their lives. Sweeping away intellectual and emotional blockages as well as feelings of victimization, tourmaline gently motivates children to take control of all aspects of their lives and acts as a catalyst as they embark on the wonderful adventure of learning to live.

Surprisingly to some, dyslexia can be a valuable learning experience that can make life more meaningful for children. When viewed from a more positive perspective, it can inspire dyslexics to live more fully and deeply. Let's not forget that some experts believe dyslexia is the reason that so many geniuses have been able to fulfill their potential. Parents would be well advised to look at the disorder from this perspective since their child's success will depend on it.

SPEECH, COMMUNICATION, AND EXPRESSION

LISA C. BERGERON

Speech is a gift, an amazing tool that we have been given. Some people find it easy to express themselves, although they may not always do so wisely. Others are naturally less talkative but more eloquent. Few people have truly developed the art of using the right word at the right time because that talent relies on complex interactions among the mind, emotions, personality, and individual insight. Given that childhood and adolescence are periods of rapid initiation to communication, they are ideal times to encourage the development of communication skills.

Blue, a color that resonates mainly with the throat chakra, is highly instrumental in developing clear and accurate speech, making blue stones the perfect tools for harmonizing or healing the throat chakra.

SUGGESTED STONES AND CRYSTAL

Chalcedony

Chalcedony is known as the stone of orators and diplomats. It can be a valuable ally in developing speaking skills since it acts on the three functions essential for good communication: listening, understanding, and expression. Chalcedony improves the memory, helps clarify ideas, confers self-confidence, and encourages tranquility and love. It can also provide support to anyone who wants to learn a foreign language. It makes us receptive to new ideas, stimulating creativity and openness to inspiration in the present moment. In addition, chalcedony helps alleviate certain respiratory problems.

Citrine

Although citrine is light yellow rather than blue, it also has a very positive effect on speech. It confers joy, warmth, and self-confidence and grants dynamic energy and courage. It helps overcome oppressive influences, thus supporting sensitive and vulnerable individuals. Furthermore, citrine stimulates the brain and reinforces the intellect, quickening our understanding. It activates creativity and new ideas and fosters self-expression. Used on its own or with a blue stone, citrine can assist anyone who wants to develop their communication skills.

Lapis Lazuli

A high-vibration stone, lapis lazuli indirectly assists communication. It stimulates self-awareness so that we can recognize our attitudes and habits, which are often unconscious, and then encourages us to change how we act and express ourselves. Lapis lazuli calms anxiety and stress as well as fosters the expression of feelings and emotions. It promotes concentration and mindfulness, awakening our inner vision and stimulating the third eye. Lapis lazuli imparts dignity and enhances inner self-knowledge, two qualities that are important for open and honest communication.

Sodalite

Sodalite can support and strengthen the building blocks of clear and accurate speech. It helps us retain our integrity while affirming our own feelings and convictions, whether through words or actions. It also helps us balance and control our emotions. It particularly assists in eliminating feelings of guilt, soothes anxiety, and bestows humility and courage.

Sodalite can release us from narrow and dogmatic ideas, granting us the capacity to observe and uncritically accept certain aspects of ourselves that have been suppressed. It contributes to reestablishing order, inspires objectivity, and stimulates logical thought while at the same time fostering intuitive perception.

If our children have trouble expressing themselves clearly, we must rely on our intuition and judgment to detect the true source of these difficulties. In this way, we can more easily determine which stones can assist them.

◆ ◆ ◆

HYPERACTIVITY, ATTENTION DEFICIT DISORDER, OR BOTH?

LISE DUSSAULT

We're hearing more and more about attention deficit disorder with or without hyperactivity (ADD and ADHD), a syndrome that particularly affects children, both boys and girls alike. The fact that the medical community has recognized ADD and ADHD has made it possible to identify the disorder earlier and more effectively. This in turn has led to a broader dialogue among stakeholders in the medical and psychosocial communities.

Symptoms

This neurological disorder is generally characterized by three main symptoms: inattention, hyperactivity, and impulsiveness. Children can display these symptoms, which may vary in severity, at home, at school, and in their day-to-day activities. Naturally, not all children with ADD are hyperactive. However, children affected by this disorder eventually not only have problems functioning normally but also experience psychological distress and suffering.

Origin of the Syndrome

Although this syndrome is known to be neurological in origin, it remains a complex disorder in which a number of factors interact. The medical community maintains that it is linked to certain anomalies in the development and function of the brain, claiming that there is an imbalance in the brain's neurotransmission process rather than an intellectual deficit.

Just as certain adaptation strategies are recommended and implemented to reduce the impact of ADD and ADHD on a child's life, certain stones can also be helpful on both physical and psychological levels. These children have to cope with numerous negative feelings, such as anger, frustration, rejection, isolation, low self-esteem, lack of confidence, intolerance, anxiety, and so on. Through their positive influence, the following stones can help them achieve more balanced behavior.

SUGGESTED STONES

Amazonite

Thanks to its soothing effect, amazonite harmonizes the nervous system by dissipating worry, fear, and anger and by balancing mood swings. It also helps balance the autonomic nervous system and internal organs and alleviates some brain disorders. It facilitates self-expression and helps develop loving communication as well.

Amber

Although actually a resin, amber is undoubtedly one of the most versatile stones in the mineral realm. For thousands of years it has been recognized and used for its powerful therapeutic benefits. Amber allows the body to rebalance and heal itself. It balances the brain hemispheres and treats and regenerates the nervous system. Emotionally, it inspires happiness, joy, peace, and confidence. It also promotes a cheerful, gentle, and easygoing nature and makes us aware of our own self-worth.

Fluorite

Many of fluorite's therapeutic properties can meet the needs of those suffering from ADD and ADHD. It enhances the activity of the nervous system, particularly that of the brain. For students, it improves concentration, facilitates understanding, and supports the learning process. In addition, acting like an energy balm, it reinforces the channeling of energies, expelling negative energy and stress and ensuring harmony within the body. In addition to these positive attributes, it also calms the emotions and promotes self-confidence.

Developing Their Strengths

Having to "manage" this syndrome personally or as it affects a family member on a day-to-day basis is a tremendous challenge. Wouldn't it be wonderful to be able to halt the cycle of recrimination, negative emotions, and lack of appreciation? It has been noted that these children are often very athletic or creative. Recognizing their talents and skills will help them form a positive and worthwhile image of themselves. The properties of certain stones will make it easier for them to develop their innate talents.

SUGGESTED STONES FOR ATHLETIC CHILDREN

Fluorite

Since fluorite acts on muscle tone and eliminates toxins, it is particularly recommended for athletes. It also confers mobility to the body and strengthens bones.

Kyanite

Kyanite reinforces the proper functioning of the cerebellum and motor nerves, thereby improving mobility and agility.

SUGGESTED STONES AND CRYSTALS FOR CREATIVE CHILDREN

Ametrine, Apatite, Pyrite

Ametrine, apatite, and pyrite, all known as sources of creativity, can be effective in treating ADD and ADHD. Pyrite enables children to draw on their abilities and potential. Ametrine improves concentration and awakens optimism and joy of living. Apatite calms hyperactivity and stimulates when overly inert. It fosters communication and self-expression. Through effective communication, we can also put situations into perspective and create a stress-free environment. Given that verbal communication helps dispel confusion, this is an important and liberating contribution.

Citrine

Citrine is an ideal stone for inspiring creativity. It stimulates the brain, reinforces the intellect, and encourages new ideas. What's more, it soothes family discord, brings joy, and elevates the spirits.

SUGGESTED STONES AND CRYSTAL FOR THE HEART

Rhodochrosite, Rhodonite, Rose Quartz

These children need to feel loved since love is a true healer. In these cases, pink stones like rose quartz, rhodochrosite, and rhodonite are indispensable. They alleviate stress, balance the emotions, boost confidence, and encourage forgiveness. In their presence, the energy of love takes its rightful place.

Return … to Nature

SUGGESTED STONES

Agate, Carnelian, Jasper, Petrified Wood, Tiger Eye

With the hectic pace of our daily lives, we sometimes forget about simple joys like playing outside. What a delight to be outdoors, fill our lungs with fresh air, and reestablish a harmonious balance. In her generosity, Mother Nature offers us a host of healing resources. Maintaining this contact strengthens our ties with nature and grounds us to the earth. This connection always confers feelings of well-being, calm, security, balance, and strength. The quality of our attention is enhanced. We can rely on a number of stones, described as "anchoring" stones, to respond to these energy needs. Agate, carnelian, jasper, petrified wood, and tiger eye are all there to assist us.

All parents want to know how to support and sustain their children throughout their development. The mineral world offers a therapeutic approach in which stones and crystals work as faithful companions and servants to assist them in this task. Through their energy reserves and the quality of their vibrations, they can benefit children suffering from any deficit.

Mother Nature "naturally" knows how to sustain and support the children of all her realms (animal, vegetable, mineral, and human). In her boundless generosity, she offers us a wealth of possibilities to discover and experience.

◆ ◆ ◆

Part 3: Adolescence and Beyond

Teenage Menstrual Cramps

Lisa C. Bergeron

Adolescent hormonal changes have so many wide-ranging effects that it may be difficult to pinpoint which reactions are directly attributable to this stage of life. Sudden mood swings, over-the-top emotional responses, excessive enthusiasm, and physical changes may surprise us as much as they do the teens who experience them. We could be tempted to arm ourselves with every mineral available to help our teens recover their balance (and make life easier for ourselves at the same time). Luckily, that isn't necessary. Working with stones and minerals has proved to be worthwhile for a number of reasons. One is their ability to spread harmony and zero in on the source of the problem while concurrently treating various related issues. For instance, the stones that will help alleviate menstrual cramps will also relieve other related problems.

Suggested Stone

Moonstone

Named after the earth's only natural satellite, moonstone's energy is closely linked to femininity and the female cycle. Like the moon, women also go through "waxing and waning" cycles that pose various emotional and physiological challenges. Moonstone regulates and softens the highs and lows triggered by women's cycles, promoting emotional and hormonal stability. It helps women rediscover their gentleness and feminine side and alleviates PMS. A calm stone, it soothes the body, reduces tension and stress, and curbs emotional overreactions. Moonstone fosters the development of emotional intelligence and helps balance feminine and masculine energies in both men and women. It is a stone of new beginnings.

Suggested Mineral

Magnesium

Magnesium has a calming effect on the body and the mind. It is a key element in relaxing the stomach, intestines, and gall bladder and alleviating muscle and menstrual cramps. Stones containing magnesium are ideal for relieving menstrual pain.

Suggested Stones Containing Magnesium

Magnesite

Magnesite conveys deep calm and reduces tension, fear, and irritation. It balances magnesium deficiencies and acts as a muscle relaxant. This stone eases the spasms that cause menstrual, intestinal, and stomach cramps. Magnesite also treats liver disorders.

By encouraging a positive attitude, it invites self-acceptance and self-love, two very important attitudes to cultivate in our teenage years. During this pivotal period in their development, young women not only experience a number of physical symptoms, but they also have to survive an emotional roller-coaster ride. Thanks to its high magnesium content, magnesite is a powerful emotional healer that will comfort and support them at the most trying times. It will also help them feel safe and supported by the Divine.

Malachite

Malachite absorbs both physical and emotional pain. It soothes and purifies the emotional and physical bodies, protecting them from unwanted energies. Malachite relieves anxiety, reduces fears, restores balance and harmony, and awakens the consciousness to unconditional love. Particularly effective in treating menstrual cramps, it also eases sexual malaise. Psychologically, this stone of transformation can intensify experiences to encourage change. Malachite brings to the surface deeply repressed emotions and the psychosomatic causes of physical disorders.

Serpentine

Serpentine transmits healing energy that acts on psychological and emotional imbalances. Not only does it attenuate mood swings, it also reduces stress, calms nervousness, and instills a serene attitude in the midst of conflict. Serpentine communicates a peaceful, gentle vibration to the emotional body and helps overcome the fear of adversity and change. This stone also relieves cramps and strengthens the stomach, intestines, bladder, and kidneys. It is an energizing stone that also promotes the absorption of calcium and magnesium.

These periods of transition are difficult for adolescents as well as their families and friends. As parents we have to be able to call on our reserves of wisdom and empathy to give our young daughters the attention and support they need to become happy and fulfilled young women in just a few short years.

◆ ◆ ◆

Anorexia

Ginette Tétreault

Anorexia, or the refusal to eat, has increased dramatically in the last twenty years. It is characterized by an obsession to lose weight and conceals a deep-seated psychosomatic disorder. When it persists, it is called anorexia nervosa. Victims of anorexia are no longer hungry or have no appetite when they do eat; food, like life, is of very little or no interest to them. This eating disorder can also be accompanied by alcohol or drug addiction.

The Consequences

Without sufficient nutritional support, excessive weight loss leads to decreased muscle mass, which in turn weakens cardiac function and arterial pressure and sometimes causes young girls to stop menstruating. When the child is very young, long-term nutritional deficiency can halt puberty growth.

The Origin

Although experts still don't understand the origin of this behavior, it is recognized that it usually starts soon after the onset of puberty, a time of profound change. This period of transition to adult life, marked by growth spurts, sexual maturation, and the nonacceptance of self, can be mentally unmanageable for some children.

SUGGESTED STONES

Always generous, Mother Earth offers us vital relief in periods of psychological and mental stress. In this case, we have a choice of three stones with strong healing powers.

Lepidolite

In times of mental distress, lepidolite, because of its lithium content, is a powerful stone for calming negative thoughts, anxieties, and fears and bringing peace and balance. It protects and helps preserve identity and independence when the desire for extreme thinness comes from outside influences.

What's more, its therapeutic virtues help overcome depression, obsessive thoughts, and all types of mental and emotional addictions. Containing iron, calcium, manganese, and magnesium, it also compensates for mineral deficiencies during this period.

Sunstone

Described as an alchemical stone, sunstone is linked to light, which enables it to establish a relationship with the regenerating power of the sun. It contains fine particles of copper and hematite that help keep a young person's feet on the ground and impart warmth, joy, good humor, and a zest for life.

This stone helps children rediscover their self-confidence and personal worth, dispelling the feelings of inferiority and pessimism that impede their recovery. New perspectives are revealed that foster hope and motivation.

Sunstone also acts as an antidepressant, transforming and banishing repressed emotions. Its healing properties stimulate the autonomic nervous system and the body's own healing powers.

Rose Quartz

Young people with anorexia have a tremendous need to be loved and suffer from low self-esteem and a strong feeling of rejection, which sap their motivation to live. Rose quartz, the quartz of compassion and a great healer, is a crucial aid in relieving this distress. It opens the heart and reinforces self-love.

Its soothing effect helps release unexpressed emotions and overcome emotional deprivation by triggering awareness of our own individual value. Its energy alleviates internalized pain, conferring the gentleness, tenderness, infinite peace, and tranquility that are essential to emotional healing.

Topaz

Topaz is a high-vibration stone that purifies the aura and soothes and recharges the body's meridians. Its vibrant, joyful, and empathetic energy promotes good health. Negativity can't survive around a topaz. It brings to light inner resources that encourage the person to let go of their anorexic behavior. A perfect emotional support, topaz facilitates digestion and restores the sense of taste.

Sustaining a young person and helping him or her abandon the malaise underlying their anorexia appear to be the vital roles stones can play here. Rain or shine, they comfort and support those who use them.

◆ ◆ ◆

Stress and Anxiety

Kristiane Roy

Today's lifestyle has dramatically changed over the decades, and much emphasis is placed on action, activity, competition, and achievement. We are, in general, overtaxed and overworked, and our children are no exception. They must perform in school, in sports, and in other extracurricular activities. There are part-time jobs for young adolescents, worries about getting into the right school, and pressure to choose a future occupation and to prepare now for an elusive future.

Social pressures and changing family dynamics add further strain. There may be a change of schools or a move to another neighborhood or city, and it is well known that more than half of families in the Western world live through the tribulations of separation and divorce.

Finding Balance

Suggested Stone

Kunzite

Kunzite balances and calms the emotions. It allows us to be focused and serene when surrounded by many distractions. This gentle stone contains lithium and is therefore effective in calming anxiety and panic attacks.

Increasing Self-Confidence

Older children and adolescents who come for treatment in lithotherapy frequently ask for help to increase self-confidence and to better express their thoughts and emotions. We all know that when they are asked "What's wrong?" a child or adolescent will undoubtedly almost always answer "nothing," even when everything is in shambles.

Coping with Stressful Events

Anxiety can manifest through repressed emotion over prolonged periods, through regular exposure to stressful and painful events that children feel helpless to deal with, or from shocks and traumatic events that they have not been able to assimilate and heal from. Children's sensitive emotional bodies may absorb much, and yet they lack the tools, as do most adults, to properly integrate what they live through.

SUGGESTED STONES

Fluorite (Green, Blue, and Mauve)

Fluorite contains calcium, an element known to aid in stress reduction. It brings awareness to repressed emotions and allows them to be gently released. It calms emotions and reduces stress as well as dispels confusion and promotes self-confidence. Green fluorite, in particular, has been known to dissipate emotional shock. Fluorite is cubic in its geometrical formation and is appreciated for its ability to bring any chaotic energy, whether emotional, mental, or physical, into order.

Rhodonite

Rhodonite is a "heart stone" that confers emotional balance and helps heal emotional wounds and self-destructive emotions. It allows us to see the solution to seemingly unsolvable problems. It diminishes emotional shock and dissipates panic attacks. Rhodonite is considered to be an essential first-aid stone during times of trial and tribulation.

From Anxiety to Serenity

The main cause of anxiety stems from preoccupation with a future that has not yet manifested. This may come from repeated negative suggestions from the media or the family and school environment, or from painful past events that are feared to occur again in the future. The mind is concerned and worried about failure, potential loss, or impending danger.

A helpful aid is to cultivate the habit of bringing the mind to the present moment and to encourage trust and faith in oneself and in life. Meditation, deep breathing exercises from the abdomen, and positive visualization can help, but it is also important to help children develop an inner strength that will help them cope with whatever life has to present. They need to know that they are not necessarily in control of what happens but they are in control of how they choose to deal with it.

SUGGESTED STONES AND CRYSTAL

Amethyst

Amethyst calms passions and violent emotions and soothes the nerves. It eases an overactive mind and encourages profound relaxation as well as a deep and peaceful sleep.

Aquamarine

Aquamarine supports meditation and relaxation practices. It works on communication blockages, granting us the courage to speak our truth and to release emotions. Aquamarine calms and clears the mind of fears and morose thoughts, thereby diminishing stress.

Lapis Lazuli

Lapis lazuli encourages mental strength, clarity, and stability and is also very supportive of meditation and relaxation. It brings to the surface traumas, repressed memories, and emotional wounds that seek to be healed, thereby diminishing causes of stress and anxiety. It stimulates communication and a healthy expression of emotions.

Note that the stones outlined in this chapter are just some of many stones in blue, green, mauve, and pink color tones that can help reduce anxiety.

◆ ◆ ◆

 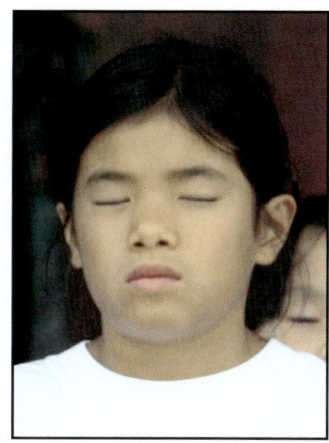

Why Meditate?

Lynda Blanchette

Caught up as we are in the hectic pace of our day-to-day lives, are we truly able to understand and meet our real needs and those of our children? Aren't we constantly bombarded by a thousand and one matters that require our attention? Don't we continually adjust our schedules to the demands of daily concerns, work, and family responsibilities?

From childhood on, we build our own conception of life based on our experiences, beliefs, and expectations. Children who don't receive the attention they need suffer deeply, and this suffering is often obvious. Once they become teenagers, the void becomes more intense, and their sensitivity to their own and others' unmet needs also increases. If they can't pause to reflect and change how they go about satisfying their needs, teens will start feeling hemmed in. That's when they begin to go along with the crowd or to rebel. Captives of confusion and illusion, they lose their way.

Meditation provides a moment for children and teens to stop, look inside themselves, and open the door to their inner being. Learning to meditate promotes concentration and encourages mental focus on a single idea or object that becomes their only focus at that precise moment in time. Like adults, they can have racing thoughts that prevent them from concentrating, calming their minds, and reaching other levels of understanding.

Stones and crystals are remarkable meditation tools for children and adolescents, who are naturally attracted to these essential companions. My daughters choose their meditation stone or crystal with care; they like to relax or meditate simply by holding them in their hands, each creating their own special moment.

SUGGESTED CRYSTALS

Celestite

A stone that imparts maternal love, celestite is perfect for children. It dissolves pain, instills love, soothes the emotions, and promotes a harmonious atmosphere during stressful times.

Citrine

Citrine, a marvelous clear yellow crystal, conveys joy of living and good humor. It fosters self-assurance, courage, and dynamism. It quickens our understanding, encourages new ideas, and stimulates creativity.

Rose Quartz

Known as the stone of unconditional love and peace, rose quartz confers gentleness and tenderness. It helps heal emotional pain and heartache, encouraging forgiveness and enabling us to acquire confidence. It also teaches us to truly love ourselves.

Creative Imagination

Meditation also helps us develop our creative imagination. Desires and fears condition our thoughts, which in turn determine our attitudes. Later, they shape what happens to us and our environment. In other words, we generally attract situations that correspond to the kind of thoughts we harbor. When we become more aware of this process, we can consciously choose thoughts of beauty, love, harmony, and success. We thus develop the potential to transform our environment and our lives. Energy flows from thought.

When children begin to learn this lesson, they become aware of the possibilities the creative imagination can offer. Once they have discovered this new inner force, they will use it as the most appropriate and intuitive way to confront the anxiety and fears they experience at this stage of their lives.

SUGGESTED STONES

Fluorite

Fluorite is the ideal stone for schoolchildren since it promotes learning. It stabilizes the emotions, imparts self-confidence, and dispels confusion. When placed in a room, it increases understanding, inspiration, and creativity.

Kyanite

Blue kyanite promotes creative expression and communication. It strengthens the memory and calms our inner images and feelings. It also eliminates anger, fears, blockages, and stress.

Why not offer our children the chance to own a stone or crystal that attracts them? They will quickly learn to rely on it as an invaluable friend.

Below I have set out a short meditation that is suitable for young and old alike. Young children will, of course, need guidance with this technique at first, but they will very quickly be able to adapt it to their needs.

The purpose of this meditation is to identify with the Light and become this Light. It raises our awareness of beauty and the joys within each and every one of us. In troubled times, it shows us the way back to our inner smile. Simple to use at any time of day and on any occasion, it can be practiced with or without the stone or crystal of your choice, depending on the circumstances.

Circulation of the Light

- Breathe very slowly and deeply.
- Feel the interdependence between your heartbeat and the rhythm of your breath.
- Focus your attention on a golden point in the center of your chest.
- Place the stone you have chosen at the level of the heart chakra or simply hold it in your hands.
- Feel your heartbeat synchronize with the energy of the golden Light circulating and expanding within you.
- Let the Light spread throughout your body while remaining in harmony with the beating of your heart.
- When you feel the Light everywhere within you, let it expand outside your body to fill the room, the building, the city, the country, the planet, and the entire universe.

- You are now one with the universe.
- You have become the Light, and this Light is without end.

Jyoti for Kids,[1] a remarkable book on meditation that also includes a CD-ROM, is now available from the same publisher. It is currently used in a number of hospitals and schools to reduce children's stress and optimize their healing and learning processes.

◆ ◆ ◆

---

[1] Sri Adi Dadi, *Jyoti for Kids: A Meditative Technique of Purification by the Light* (CD-ROM) (Montreal: Paume de Saint-Germain Publishing, 2009).

PART 4: CRYSTAL THERAPY FOR SPECIFIC CIRCUMSTANCES

STONES: PILLARS OF THE ADOPTION PROCESS

KLAIRE D. ROY

The Extraordinary Adventure of Adoption

The arrival of a new child is always an extraordinary adventure. Sleepless nights and dirty diapers are a small price to pay for building the inextricable bond between parent and child. This bond, surely one of the strongest in the world, is usually forged from the moment of conception. On one hand, parents experience an indescribable feeling of love, and on the other, the child feels secure in this love. However, the scenario is slightly different for adoptive parents. They have had time to prepare during the months leading up to the arrival of this eagerly awaited addition to their family, of course, but what about the child?

Meeting the Challenge

Once the feelings of anxiety about the first meeting have dissipated, the happy parents then have to deal with the demands of the adoption itself, which poses a dual challenge. Overnight they've become responsible for a new life that will shape them as parents, and in many cases they also have to adjust to a child who has come from the other side of the globe, who speaks a foreign language and doesn't understand theirs.

The period of adaptation is actually a time of intensive bonding. The key to smoothing this transition is the love the child receives, which unlocks the door and sustains both parents and child. Often having endured hardships and starved for love, these children may have emotional problems that the new parents will have to heal in order to restore balance to the lives of these new members of the family.

These children may express the insecurity and lack of love they've experienced since birth in a variety of ways. Some have frequent nightmares and inexplicable fears, and they may even refuse to let one of the parents approach them. Some are hypersensitive, betraying their fear of abandonment. Others virtually shut down, withdrawing into an inner world that is inaccessible to the parents. How can we give these children the all-encompassing and effective help they need?

The Support of Stone and Crystal Elixirs

My own experience has clearly shown me that stones can have a remarkable effect. I am delighted to say that my daughter, who was adopted nearly six years ago, is a constant source of happiness and joy in our home. To help her adapt to a wide range of situations, we have used stone elixirs on a regular basis. I call them "miracles in a bottle." We're now preparing to adopt a second child in a few weeks, this time a little boy, who will be another sunny addition to our lives. And so I'm setting off on this trip with a selection of stone and crystal elixirs in my luggage.

SUGGESTED STONES AND CRYSTALS ELIXIRS

Amethyst

I'll slip into my suitcase an amethyst elixir that can calm fears and improve the quality of sleep. Our child will thus feel safe and sleep more peacefully. Amethyst will also give him the strength he needs to adapt with ease to the many changes his new life will bring: a new language, a new country, new faces, and so on.

Citrine

Citrine elixir will also be accompanying us on this trip. It will add a dose of light-heartedness to the occasion by minimizing our son's feelings of anger and anxiety. His digestive system will also benefit from the support citrine can provide and help him adjust to his new diet.

Emerald

And, of course, I mustn't forget emerald elixir, which promotes the development of a healthy immune system. If the child catches a cold, which often happens, emerald elixir will act as the perfect complement to treat his lungs and bronchial tubes. Another of the emerald's properties is that it instills a strong sense of being loved and recognized, even in the most difficult times. Our son will therefore feel loved and protected in his new environment.

Rose Quartz

For me, rose quartz elixir is indispensable to help ease and heal the sorrows our child has experienced. It will soothe his deep-seated feeling of abandonment, allowing him to feel loved and comforted.

The parental adventure, an experience of the utmost importance, takes careful planning. Stones, crystals, and their elixirs, which have the capacity to sustain us, are strong pillars that we can rely on to support us throughout the various stages of the adoption process. I have to admit that I regularly use stone and crystal elixirs to help me fulfill the wondrous yet sensitive role of motherhood. I use amethyst elixir to cultivate patience and citrine elixir to be able to cool-headedly cope with the many unsettling situations that arise on a day-to-day basis. As for rose quartz elixir, which I adore, it helps me feel loved at all times, while emerald elixir ensures better health. What more could I ask for?

◆ ◆ ◆

ALLERGIES

JACQUELINE D. SYLVAIN

An allergy is the body's inappropriate response to a product, substance, or allergen it identifies as harmful. The organism then sets off a number of processes to eliminate what it considers to be a potentially dangerous enemy.

The physical symptoms experienced are largely a result of this battle between the immune system and the aggressor, which is usually found in food (nuts, dairy products, eggs, seafood), the environment (pollen, dust, mold, animal dander), or chemical and organic products (medication, latex, insect venom).

Symptoms

When an allergy appears, the first step is to reduce the immune system's responses. Once that has been done, it's important to discover the cause to make sure the allergy doesn't lead to more serious problems. The most common allergy symptoms are a runny nose, sneezing, congestion, itchy eyes, persistent cough, and even asthma. Recurrent ear infections, eczema, digestive disorders, or skin rashes of unknown origin are also common.

Although there is a broad range of medication available for allergy relief, the problem remains. Anything that boosts the body's natural immune system is an effective weapon in the war against allergies.

SUGGESTED STONES

Aquamarine

According to ancient legend, aquamarine helps its owner distinguish friend from foe. That may be why it tops the list when it comes to treating allergies, since it significantly reduces the immune system's excessive response to the invasion. In fact, the enemy that is causing the allergy is often a problem within ourselves.

Aquamarine is the first stone to be included in our antiallergy kit because of its ability to alleviate the immune system's extreme response and thus combat the allergy. Known as "the stone of courage," it neutralizes the effect of environmental pollutants and increases resistance. This stone also helps reduce stress and enables us to recognize its underlying emotional causes.

Danburite

Danburite is endowed with a very pure vibration that helps relieve allergies and chronic disorders. It also has a strong detoxifying action.

Emerald

Emerald is the second most important stone for treating allergies. It reinforces the immune system and promotes the healing of infectious diseases. Anything that can help enhance the body's natural defenses is effective against allergic reactions.

On a spiritual level, emerald promotes friendship, love, partnership, and unity, stimulating growth and inner harmony. As indicated above, the physical symptoms of allergies are the result of the struggle between the body's immune system and the aggressor or invading agent. This stone also helps us recover our balance and cope with problems that come our way.

Lepidolite

An effective purification tool, lepidolite emits gentle vibrations that soothe allergies. It reinforces the immune system, calms emotional stress, and helps us recover and reestablish balance.

SUGGESTED STONES AND CRYSTAL FOR SERIOUS ALLERGIES

Amber

Amber, which is actually fossilized tree resin, is a powerful cleanser and healer. It reduces allergic reactions and promotes the revitalization of the mucous membranes and other tissues. In cases of infection, amber acts as a natural antibiotic.

Amber can be a valuable aid when allergies become more severe and cause more serious respiratory problems. Its properties enhance the power of self-healing.

Apophyllite

Apophyllite is also useful for particularly severe allergies that lead to serious respiratory problems. It helps us achieve a state of deep relaxation, which enables the healing energy to act. Any allergy puts enormous stress on the human organism. Placed on the throat at the level of the thymus, apophyllite relieves asthma attacks. In elixir form, a few drops under the tongue are enough to neutralize the allergy and regenerate the mucous membranes.

Rutilated Quartz

Rutilated quartz has all the qualities necessary to combat chronic infections and can help heal the respiratory tract.

SUGGESTED STONES FOR SKIN ALLERGIES

Aventurine

Aventurine has an anti-inflammatory effect on skin rashes and eye inflammation. It confers a general sense of well-being and calms the emotions.

Rhodochrosite

Rhodochrosite helps absorb and reduce the effect of irritants that cause allergic skin reactions. This stone is also very effective in relieving respiratory disorders. In addition, it lessens the severity of migraines, which often occur during these periods.

Suggested Stone and Crystals for Digestive Disorders

Some intestinal and digestive disorders are allergies triggered by the ingestion of allergens.

Amber

Amber is excellent in these cases since it absorbs pain and negative energy, encouraging the body to rebalance and heal itself.

Amethyst

Amethyst is very effective in relieving intestinal disorders. It also has a beneficial effect on skin conditions and swelling. It purifies the blood and strengthens the immune system.

Citrine

Citrine is a true friend to the digestive system, and as such provides strong support in relieving stomach and intestinal disorders.

It's important to live a healthy and well-balanced life. Despite the troubles we may encounter, Mother Nature offers us everything we need to achieve the quality of life we seek. We simply have to choose the right stone or crystal that will help us attain this state of well-being. And the good news is that allergies sometimes vanish on their own given time.

◆◆◆

Asthma

Jacqueline D. Sylvain

Asthma, which means "panting" in Greek, is a disease that affects the pulmonary system. It is caused by spasmodic contraction of the bronchial tubes and bronchioles, which makes breathing difficult. It can be aggravated by the accumulation of a significant buildup of mucus in the bronchial tubes and bronchioles. Asthma attacks can be triggered by a number of factors.

There are three types of asthma: chronic asthma, allergic asthma, and exercise-induced asthma.

Chronic Asthma

Chronic asthma is characterized by permanent hyperactivity of the bronchial tubes, which are infrequently exposed to outside agents. This form of asthma is caused by an autoimmune reaction in which the immune system attacks its own tissues, the bronchial tubes. The inflammation slowly and gradually becomes chronic.

Allergic Asthma

Allergic asthma is caused by an outside agent (allergen) that triggers an excessive reaction of the bronchial tubes. An attack starts with a sudden and rapidly progressing obstruction of the bronchial airways. This form of asthma is the most aggressive in the short term and can be fatal.

Exercise-induced Asthma

Exercise-induced asthma occurs during a physical effort that "traumatizes" the bronchial tubes and prevents them from functioning correctly. Environmental factors like cold and wind aggravate this form of asthma. Intense stress or emotions can also cause an attack since they accelerate the heartbeat, causing hyperventilation.

Childhood Asthma

Pediatric asthma is the most common chronic childhood disease, affecting more than twenty-five percent of children. Accordingly, asthma symptoms are a source of serious concern for many parents. When their child's asthma spins out of control, they have to rush to the emergency room, and the child may frequently have to be hospitalized.

Asthma medication has many undesirable side effects, including digestive, skin, and growth problems. That's why it's important to work with a natural approach that poses no threat to health and will arm the child with a balanced defense system.

SUGGESTED STONES AND CRYSTAL

Apophyllite

Apophyllite, particularly green apophyllite, is one of the stones most frequently recommended for asthma. It encourages a state of deep relaxation, which is crucial for coping with this disease. It also acts quickly and effectively to soothe respiratory problems. Simply placing the stone on the throat will ease the asthma attack.

In psychological terms, asthma is essentially related to feelings of suffocation, anxiety and insecurity. Apophyllite helps alleviate fears and oppressive feelings and releases repressed emotions. In addition, it encourages us to overcome problems and insecurities.

Magnetite

Magnetite is indispensable against asthma and immune deficiencies. It stimulates sluggish organs and calms hyperactive ones. It provides the healing energy needed for recovery. It also regenerates cells and restores their natural magnetism.

Malachite

Malachite is another stone that is effective in treating asthma. It fortifies the immune system and facilitates deep emotional healing. It draws out suppressed emotions and restores the ability to breathe deeply.

Pyrite

Pyrite is beneficial to the lungs, alleviating asthma and bronchitis by clearing the bronchial tubes. It is an excellent energy shield, blocking out negative energy, including that associated with infectious diseases. It assists in understanding the causes of the disease so that the necessary changes can be made.

Rhodochrosite

Rhodochrosite filters irritants, relieving asthma and respiratory problems. It alleviates emotional turmoil and irrational fears and also works very well with malachite.

Rutilated Quartz

Rutilated quartz brings beneficial vitality to people suffering from chronic pulmonary disease. It stimulates the regeneration of all cells and treats respiratory problems, including chronic bronchitis.

Psychologically, rutilated quartz facilitates transition and change and supports the development of new ways of life. From an emotional perspective, it soothes the fears and anxieties of asthma victims and restores hope.

It should be understood that asthma sufferers must be able to express what is suffocating them, to occupy their own space, and thus to take their rightful place in life. They need to be able to open their hearts and work according to the integration process that corresponds to their genuine needs. Each time they inhale and exhale, they must let go and let the current of life flow through them.

◆ ◆ ◆

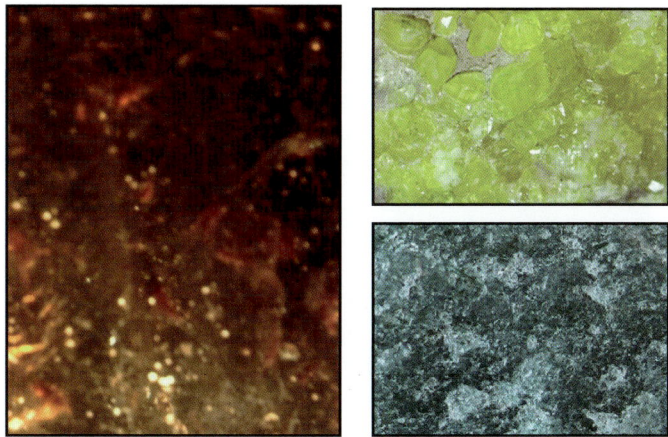

ECZEMA

GINETTE TÉTREAULT

Eczema or atopic dermatitis affects children and infants from birth in the form of red blotches or patches and dry skin that causes itchiness. It most frequently appears on the cheeks, hands, behind the knees, in the elbow crease, and on the ankles.

How You Can Help Your Child

Research has shown that eczema is linked to psychological and emotional problems. Fortunately, the earth can offer us the vibratory and healing power of amber, aventurine, and sulfur.

SUGGESTED STONES

Amber

Known as a powerful healer, amber draws disease from the body, absorbing the pain and stimulating the body to rediscover its balance. It promotes the regeneration of the nervous system, which is often responsible for skin conditions. Its resonance brings joy to children who wear it, calming their frequent fears. It acquires an electric charge by friction and produces negative ions that promote the circulation of energy to benefit the entire body.

Amber thus contributes to healing wounds by promoting tissue revitalization and boosting the immune system. Since its therapeutic effect is gradual and long-term, it is recommended that the stone first be worn by the mother before using it to treat the child. In this way, the amber will be imbued with positive vibrations, joy, tenderness, and maternal love.

Aventurine

The beneficial effects of green stones that raise the body's vibrations above the illness are well recognized in lithotherapy. Aventurine thus has anti-inflammatory properties that eliminate rashes stemming from, among other things, an imbalanced nervous system. Thanks to its chemical composition, silicon dioxide, it can help strengthen the skin and connective tissues, accelerating the healing of patches of dermatitis.

Through its affinity with the center of the heart, which makes it an ideal remedy for ailments of this organ, it can alleviate emotional stress in children. It acts as a soothing balm on the emotional body, compensating for a lack of love and understanding in the family.

Sulfur

Created by volcanoes, sulfur is a powerful ally and an excellent stone for any skin condition. Containing keratin, it promotes the well-being of the superficial cells of the epidermis. If you dip your hands in a sulfur spring, you'll notice that they're immediately rehydrated. Sulfur thus actively alleviates eczema-related problems thanks to its antiallergenic properties and has a universal desensitizing effect. It can absorb negative energies and emotions that are harmful to children.

However, care should be taken when using sulfur since it contains toxic agents. We recommend external use only, simply placing the stone on the body. If this proves to be irritating, a thin piece of cloth should be placed between the skin and the stone to relieve the inflammation. Although stones and stone elixirs are harmless in themselves, it is important to use them carefully and consistently to ensure their effectiveness.

◆ ◆ ◆

DIGESTIVE PROBLEMS

LISA C. BERGERON

Digestion is a key physiological function. The digestive system is responsible for our ability or inability to absorb the nutrients that are essential for the body to function properly and eliminate harmful substances. Digestive problems can manifest in adults and children alike as constipation, intestinal cramps, reflux, diarrhea, vomiting, hiccups, gas, bloating, and heartburn.

Digestive disorders may result simply from eating something that's hard to digest, overeating, or lack of certain vitamins or minerals. And similarly to many other disorders, digestive problems may go hand in hand with emotions such as stress, fear, sorrow, and anger.

SUGGESTED STONES, CRYSTAL, AND ELIXIR

Amber

Amber imparts heat that soothes and warms the emotional and physical body. It establishes an inner stability that enables both body and spirit to consistently fulfill their functions without fatiguing. This tree resin, which is over forty million years old, is extremely beneficial for alleviating digestive problems. Its ability to absorb negative energy naturally encourages the body to begin a process of reharmonization, thus making amber an exceptional purifier and healer. In fact, amber has been called a natural antibiotic. It relieves stress and treats imbalances in the stomach, liver, and spleen.

Amber is also helpful for many aspects of childhood development. For instance, it is excellent for teething pains. When using a piece of amber or a necklace for a baby, the mother should carry or wear it first for a few days so that the stone can absorb her energy in resonance with that of her child.

Chrysocolla

Chrysocolla purifies, calms, and reenergizes all the chakras while harmonizing them with their divine nature. It can detoxify the liver, colon, and kidneys, reinforcing their natural healing power. It also treats disorders of the digestive tract, revitalizes the pancreas, regulates the blood, and alleviates muscle cramps.

Chrysoprase (Elixir)

Chrysoprase elixir is particularly effective for stress-related stomach problems. It is known for soothing digestive disorders, stimulating the liver and detoxifying the body. It helps establish hormonal balance and improves the absorption of vitamin C, which is required for healthy tendons, bones, teeth, blood vessels, and muscles. Chrysoprase elixir plays an important role in developing tissue resistance and joint flexibility. It also promotes elimination and strengthens the immune system.

Citrine

Citrine is an excellent support for the digestive system. It stimulates digestion and the proper functioning of the stomach, spleen, and pancreas, and also relieves constipation. By generating inner heat within the body, it relaxes physical contractions and can moderate our psychological reactions. Citrine grants confidence, courage, energy, and joy of living. It also helps overcome depression and oppressive influences. Mentally, citrine helps us digest events as they occur. It encourages children to face who they are and understand their reactions to certain situations.

Most children tend to display somatic tendencies: they may experience their feelings and moods as physical symptoms because they lack the vocabulary to express themselves. That's why we need to pay close attention to what their bodies are telling us.

◆ ◆ ◆

PHYSICAL TRAUMA

JOANI GAGNON

Our children are budding explorers who set off on new adventures each and every day. Unfortunately, they can sometimes hurt themselves and run home crying, or even be struck by a panic attack. Once we've applied our knowledge and our first-aid kit, how can we as parents use stones and crystals to help heal our children's wounds?

Emotional Wounds

SUGGESTED STONES AND CRYSTAL

Amethyst, Aragonite, Blue Tourmaline, Kyanite, Sodalite, Sunstone

As important as it is to heal the physical body, it's equally important to heal emotional wounds. When our children experience trauma, it's crucial to encourage them to express their feelings. Whether they've fallen off their bike or out of a tree, they are sure to have felt fear. Giving them a blue tourmaline, sunstone, or sodalite to hold in their hands or slip into a pocket can help dispel this fear, lessen the shock, reduce the impact of the trauma, and alleviate their anxiety. When our children are feeling angry (if a friend has pushed them, for instance), amethyst, aragonite, and kyanite can calm and banish their resentment, anger, and frustration. That's why it is imperative that we listen to our offspring and help them blithely set off again on the adventure of life.

Sunburn and Other Burns

SUGGESTED STONES AND CRYSTAL

Chrysocolla, Lapis Lazuli, Rose Quartz, Sodalite, Turquoise

Clear blue skies, warm temperatures, outdoor games, and family outings are all part of the joys of summer. Unfortunately, once evening rolls around and our children take a look in the mirror, they may see the unwelcome effects of sunburn. In the case of a simple burn from the sun or other sources, placing blue stones, like lapis lazuli, sodalite, or turquoise, can ease the pain. As the coldest color in the spectrum, blue is soothing and refreshing. After that, chrysocolla or rose quartz can help regenerate the skin. Naturally, frequent applications of sunscreen are also recommended. After all, isn't an ounce of prevention worth a pound of cure?

Bruising and Nosebleeds

SUGGESTED STONES

Andradite Garnet, Aquamarine, Blue Chalcedony, Calcite, Hematite, Larimar, Pyrite, Rhodochrosite

Bruising and nosebleeds occur when trauma damages and breaks open the blood vessels in the skin, and they are often the result of weak capillaries, which in most cases are in turn due to a lack of calcium or vitamin C. Andradite is useful for calcium deficiency since it helps produce hemoglobin and contributes to calcium absorption, thus increasing capillary resistance and reducing the risk of bleeding.

Composed of iron oxide—oxygen and iron—hematite is also effective in preventing or arresting bleeding since it works on the bloodstream at a number of levels. Not only does it promote the development of red blood cells by providing iron and oxygen, it can also eradicate signs of anemia over the long term. By stimulating iron absorption, hematite also helps strengthen the capillaries.

If your child has a serious nosebleed, stones like calcite and blue chalcedony can help stimulate blood clotting. An aquamarine or larimar placed on the throat is also effective. In addition, these blue stones can help your children express their fears and calm down.

Blue chalcedony, pyrite, and rhodochrosite are also beneficial for the bloodstream. They stimulate blood circulation, strengthen the blood vessels, and restore their elasticity. It should be noted that these lesions can occur more frequently if your child lacks vitamin K.

Scrapes, Scratches, and Surface Wounds

SUGGESTED STONES AND CRYSTAL

Amber, Apophyllite, Aventurine, Emerald, Jade, Malachite, Rose Quartz, Selenite

The scrapes, scratches, and surface wounds that are the most common childhood injuries should always be properly cleaned and disinfected. Once that has been taken care of, placing a green stone close to the wound can be helpful since it has regenerating and soothing properties. Apophyllite and malachite are ideal for these types of injuries. As well, stones containing silicon, such as aventurine, jade, rose quartz, and selenite, play a major role in cell division and accelerate the healing process. However, if the wound becomes infected, it should be cleansed as often as possible, and stones with strong purifying properties such as emerald are recommended. If the infection persists, amber, which acts as a natural antibiotic, may be used over the long term.

Fractures

SUGGESTED STONES

Aragonite, Blue Lace Agate, Green Apatite, Green Calcite, Magnetite

Our children can often be victims of more serious trauma, such as fractures or broken bones. But not to worry: wearing a cast won't usually keep them housebound for long. Stones like blue lace agate, green apatite, green calcite, and magnetite, which regenerate cells, will help their bones absorb calcium. Aragonite improves the absorption of vitamin D and calcium, which plays an important role in bone and tissue development.

Muscle and Joint Inflammation

SUGGESTED STONES AND MINERAL

Azurite-Malachite, Blue Fluorite, Copper, Magnetite, Rhodonite

Muscle or joint inflammation often occurs following a sudden violent shock to the muscles or joints, including the stretching of tendons or ligaments. Pain, redness, heat, and swelling are all signs of inflammation. Applying an ice pack to the painful area for five to ten minutes immediately after the injury is recommended in these situations. After that, azurite-malachite, blue fluorite, magnetite, rhodonite, and copper can all help alleviate joint inflammation.

Insect Bites

SUGGESTED STONES

Azurite, Lapis Lazuli, Rhodonite

And then there's the dreaded insect bite. Although we take countless and sometimes even ridiculous precautions to ward off insects, there's always one that manages to slip under our radar and zero in on our child. Applying green clay to the affected area will draw out the stinger. Azurite, lapis lazuli, or rhodonite can then be used to reduce the swelling and itching.

Various Muscle, Ligament, and Joint Injuries

SUGGESTED STONES AND CRYSTAL

Blue Spinel, Citrine, Dioptase, Fluorite, Labradorite, Malachite, Moss Agate, Rutile

Children can also suffer from muscle, ligament, and joint injuries, including pulled muscles, tendons, and ligaments. In case you've forgotten your anatomy, a tendon connects a muscle to a bone, while a ligament connects muscles to each other.

Stretched ligaments and muscles, which are more commonly called sprains or strains, may be mild or severe. They occur more often than we may think, affecting children and adults alike. A mild sprain is caused by a sudden, brisk movement of the joint that overstretches the ligaments. In a more severe sprain, the tissues are also torn. Green stones are effective in relieving stretched muscles and sprains since the color green calms, neutralizes, and balances. Moss agate, malachite, and dioptase can help ease inflammation and pain.

However, with this type of trauma it's also important to strengthen the muscles. Some physical exercises are a must, as are certain stones and crystals. For example, thanks to their chemical composition, citrine, fluorite, labradorite, rutile, and blue spinel help fortify the muscles. Sprains also include partial or full dislocation when the bones in the joint become loose or slip completely out of their sockets, stretching all the muscles and ligaments that initially held them in place. A specialist should be consulted in these cases.

Muscle Tension and Cramps

SUGGESTED STONES

Amazonite, Charoite, Magnesite, Selenite

Most of us have been jolted awake at night or stricken down on the playing field by a piercing muscle cramp. Muscle cramps and contractions are caused by an imbalance of minerals such as sodium, calcium, magnesium, and potassium. Fortunately for us, amazonite, charoite, magnesite, and selenite contain these minerals and can bring relief when placed on the affected area. Because of their antispasmodic properties, they help relax the muscles and alleviate pain. Massaging and stretching the muscle so that it can draw support from the minerals contained in our blood is also recommended.

We can bandage or otherwise treat our children's minor injuries knowing they'll heal over time, but we also need to sustain them emotionally and enable them to express their feelings. Stones and crystals can help them understand and accept the emotions underlying a quarrel, a fall, or any minor accident. Why not let them choose a stone or crystal that can accompany them all day long and serve as their own personal treasure?

◆ ◆ ◆

THE IMMUNE SYSTEM

JACQUELINE D. SYLVAIN

The immune system, which is the body's defense mechanism, protects us from external agents like bacteria, viruses, fungi, and so on. If our immune system isn't healthy, a simple scratch can be fatal. This system is also impacted by our emotions (anger, self-destruction, hate) and by our positive and negative thoughts, which may strengthen or weaken it. That's why it is vital to make sure that our immune system functions as well as possible by giving it the opportunity to effectively combat aggressors. A well-balanced lifestyle, healthy diet, exercise, and the use of stones and crystals will enable us to improve the body's immune responses and stay healthy.

In today's industrial society it is wise to develop a strong immune system early in life, and good habits and prevention are the best ways to do so. From the moment they're born, our children are exposed to a toxic environment where electromagnetic waves, pesticides, pollution, and countless other negative influences abound. Their exposure affects their immune systems and can lead to colds, allergies, ear infections, eczema, and asthma, distressing both parents and children.

SUGGESTED STONES AND CRYSTALS

Amber

A natural antibiotic, amber is useful for infection and fever. Applied as an elixir on the base of the neck, the inside of the wrists, or the solar plexus, it can provide effective relief for infants and children. A powerful healer and purifier, it imparts vitality, absorbs pain and negativity, and promotes the healing of wounds. It also enables the body to recover its balance and heal itself. Psychologically, amber imparts joy and self-confidence. It also relieves stress and emotional turmoil. Furthermore, its warm, luminous energies are highly suitable for children.

Amethyst

Amethyst reinforces the cleansing and eliminating organs as well as the immune system. It also regulates intestinal flora. Considered a natural tranquillizer, it relieves stress and nervous-system disorders, promoting tranquility. It helps combat insomnia and protects against nightmares by inducing restful sleep. Amethyst balances our emotional highs and lows and is a valuable aid in eliminating anger, fear, and anxiety.

Ametrine

Ametrine, which is composed of amethyst and citrine, helps strengthen the immune system. It combines the qualities of amethyst and citrine to help dissolve negativity, bodily toxins, and fatigue. It also treats depression.

Aquamarine

Aquamarine is an excellent choice for alleviating excessive immune reactions such as asthma and allergies. Its power is enhanced when it is used in conjunction with emerald. Using these stones together helps combat throat and sinus infections and cure colds quickly. In elixir form, they act as a "flu buster."

Known as a stone of courage, aquamarine transmits the strength to meet any challenge. It brings soothing energies that help diminish stress and calm the mind.

Calcite

Calcite, particularly green calcite, fortifies the immune system. It promotes the absorption of calcium, thereby stimulating growth in young children. It also relieves emotional stress and confers peace of mind. Calcite brings stability, increases self-confidence, and helps overcome all obstacles.

Emerald

A powerful ally of the immune system, emerald can help us remain in good health when our body's natural defenses are low. It regenerates our physical body and contributes to rapid recovery from infectious diseases. This stone also ensures physical, emotional, and mental balance, erases negativity, and bestows a positive attitude. Emerald opens the heart and enhances our ability to enjoy life to the fullest.

Lepidolite

Lepidolite is a strong purification tool that dispels all negativity. A calming stone, it relieves sleep disorders. It is also extremely useful in eliminating stress, depression, and mood swings.

Aqua Aura Quartz, Aragonite, Chalcedony, Chiastolite, Fluorite

Other stones also sustain the immune system. Aqua aura quartz strengthens the thymus; aragonite combats anger and stress; chalcedony dissolves negative thoughts and feelings; chiastolite balances, protects, and stabilizes the emotions; and fluorite is a powerful antiviral agent, especially in elixir form.

All these stones are invaluable partners that can help our children develop into healthy adults. These are natural tools that will provide the balance and harmony children need to face the future.

◆ ◆ ◆

A Dying Child

Nancy Bédard

The death of a child is unquestionably the most painful experience a parent can face. It is practically impossible to deal with the heartbreak when such a death occurs without warning. However, when death is expected, there are a number of options open to parents to help them meet this challenge, learn from it, and accompany their child through this ordeal.

We know very well that each of us has our own destiny and soul. However, we soon forget this lesson when our children are involved. From the day of their birth, a visceral bond is established between us. From that time on, we devote ourselves to meeting all their needs and protecting them, and we make their lives our own. But when a diagnosis of probable death is pronounced, we are devastated; life loses its meaning and becomes intolerable. Yet despite our sorrow, we have to accept the situation and allow this life, to which we gave birth, its own existence.

How Does a Dying Child Experience Approaching Death?

Until the age of three, children don't understand what death means. They react to the pain and discomfort of illness. Highly sensitive to their parents' distress on their behalf, they suffer from feeling their parents' sorrow and being separated from them. From ages three to five, children are more concerned by the absence of their parents than by the imminence of death. They are afraid of being abandoned and left alone. They tap into the feelings of sadness and anger around them and may interpret them as rejection.

The attitude of school-age children, who understand that death is inevitable and irreversible, is strongly influenced by those around them. Deeply distressed by their parents' sorrow, they feel guilty about being sick and think they are responsible for their parents' pain.

As for teenagers, they clearly understand the challenges death brings. They suffer intensely from this knowledge as they have to come to terms with all the losses it brings. At a time in their lives when they are desperately seeking independence and autonomy, they have to accept that there is no future, live with the premature breakdown of their bodies, and accept that they once again have to depend on others.

Ironically, despite the physical pain and because they are not yet weighed down by beliefs or fears, dying children instinctively know that there is more to life than the simple physical body, that death is not an end but a continuation. Children's understanding of death far exceeds that of adults; certain aspects of their inner experience are beyond adult logic. Thanks to their extraordinary openness to the wholeness of life and their innocence, purity, and light-heartedness, they are far more serene in the face of death than we would expect.

As death approaches, children intermittently live between two worlds. They go through periods of silence and withdrawal, which are understandable since they take refuge in their inner world. The inability of those around them to understand that world makes them feel helpless, sad, and sometime even lonelier. Of course, they suffer from the effects of their illness, but the nonacceptance and silence of their parents, friends, and family add to their distress and have a huge impact on their final moments.

Through their wisdom and service to humans, stones bring support and comfort to the intense suffering of these children and their families. They help raise the consciousness of the dying and those around them, preparing them for the upcoming transition. They are an invaluable aid to a child in this process of separation.

SUGGESTED STONES AND CRYSTAL

Kunzite, Pink Calcite, Rhodochrosite, Rhodonite, Rose Quartz, Sugilite

Stones that resonate to the color pink, such as kunzite, pink calcite, rhodochrosite, rhodonite, rose quartz, and sugilite (magenta), convey calm and gentleness, soothing anger and fear. They provide comfort and security, reassuring children of the important place they occupy in their parents' hearts. These stones transmit the love of self and of others, alleviating heartache and the need for love. Their gentle energy promotes forgiveness, encouraging children to let go of those they love and bringing them closer to their inner world.

Aquamarine, Kyanite

Aquamarine and kyanite are also recommended in these circumstances. Their properties enable them to soothe the mental body and to calm the fears, worries, frustrations, anger, and stress caused by the unspoken thoughts of those surrounding the dying child. By purifying the throat chakra, these stones encourage children to express themselves and communicate their inner truth, which is sometimes ignored by the parents because of their profound attachment to their child. These stones also confer lightheartedness and serenity. Kyanite aligns the subtle bodies and purifies the body's various energies. It calms inner vision and emotions, thus easing the transition through the dying process.

Carnelian

Carnelian is also recommended in these cases. Known as an anchoring stone, it grants stability and the courage to face painful situations. Clearing mental confusion, it helps us believe in and stand up for ourselves. In addition, it banishes envy, rage, and resentment—emotions any parents feel when they are forced to face the imminent death of their child. Its energetic vibration facilitates acceptance of the cycle of life, reminding us that everything that is born must eventually die and thus eliminating the fear of death. It also strengthens our connection with the higher self, the force that animates and guides each and every one of us. Wearing a carnelian helps maintain high energy levels. It is recommended for parents except in cases of cancer, since emanations from orange and red increase the vitality of all the cells, including cancerous ones.

Charoite

Charoite is a sustaining and reassuring guide for frightened children because it aligns and unifies them with their soul. It also helps them adapt to the changes that approaching death brings. It alleviates feelings of insecurity and boosts the self-confidence that children so badly need at this difficult time, enabling them to remain balanced. Charoite's vibrations promote deep emotional healing that helps sick children accept the present moment as perfect despite the chaos around them.

Preparing Your Child for Death

Although comforting their child as death approaches is an intensely traumatic experience for all parents, it is also the greatest proof of love they can give. It is crucial for them to be at the bedside of their child who is preparing to leave this world. Their support and patience at this time are invaluable.

Dying children need to be able to talk about death; to be listened to, supported, and understood through all the suffering and pain. They must be able to ask questions without worrying about being judged. Through establishing frank, direct, and open dialogue, parents can allow their children to express their fears, desires, and disappointments and break down the walls of solitude on both sides. In this way, they are all better armed to understand each other and prepare for the inevitable.

Dying children recognize their inner voice and know that this voice is indeed alive and will guide them through the moment when their body ceases to house their soul. But they don't always receive the help they need from those around them, which frightens them and is contrary to what they desire. Death comes when the soul decides to leave the physical body in order to continue its evolution in another form. Stones and crystals are excellent allies for the soul in this process of withdrawal since they help the personality concentrate on what it needs to separate itself from the body that envelops it.

SUGGESTED STONE AND CRYSTAL

Amethyst

Amethyst, a powerful and protective stone with a high vibration, purifies the emotions by reducing anxiety and stress. Highly beneficial for those reaching the end of their lives, it makes it easier for them to let go and calms and refines their thoughts, bringing inner peace and understanding of what is taking place. Amethyst also harmonizes and raises internal energies, which facilitates children's connection with their innermost nature. It dispels anger, sorrow, and fear, thus easing their transition toward death.

Danburite

Danburite is a spiritual stone that can help us access our inner spiritual path. It harmonizes the body's energies and cleanses it of all blockages, assisting in the preparation for death. Danburite aligns the heart chakra with the higher chakras, helping us to connect to divine consciousness. Highly recommended for teenagers, it promotes in-depth change and encourages them to focus their attention on the soul rather than on the personality and on any regrets they may have, enabling them to accept the past and leave it behind. Placed by the bedside, danburite accompanies the dying through their voyage beyond death, allowing them to make a conscious spiritual transition.

Death is a natural process that can be a simple transition. However, our ties, beliefs, and fears complicate this passage, making it more difficult. Stones and crystals can provide us with indispensable, invaluable assistance that will guide us through this important stage of life. It is only once we are at the gates of death, freed from all our roles, that we can understand that we are all equal beings who have come to earth to learn from everyone around us.

◆ ◆ ◆

Pediatric Cancer

Kristiane Roy

Despite medical treatments that do prolong life and cure many, a diagnosis of pediatric cancer is often received with fear and dread. Children, parents, and other family members experience a complex array of social, emotional, economic, and familial stresses.

Young cancer patients must maintain a delicate balance between managing their body's response to the illness and their emotions. A child who has successfully adjusted to the illness has had to contend with an array of fears, pressures, and frustrations. A critical stage of growth can be disrupted during this challenging and stressful journey. Children may feel cut off from their peers, feeding feelings of isolation, while enduring uncomfortable medical interventions.

We in no way endorse the use of stones as a replacement for traditional medicine, but rather as a beneficial support to ongoing care. Children have a natural affinity to the mineral kingdom, whose array of vibrant colors, textures, and forms naturally bring a sense of joy and pleasure into their lives. We have seen children carry them in their pockets as companions, sleep with them under their pillows for reassurance, and hold them in their hands as an anchor during times of stress.

GREEN STONES AND OTHER MINERALS

Green stones are the most beneficial for cancer as this color is known to purify and balance. It is the color that elevates the body above the vibration of illness; it is widely known that hospitals have fewer incidents of infection where there is a predominance of green in the environment.

Aventurine

Aventurine absorbs electromagnetic pollution and protects against environmental toxins while lending support during the treatment of malignant diseases.

Green Calcite and Malachite

Green calcite and malachite absorb, purify, balance, and soothe.

Chlorite Quartz

Chlorite quartz supports the treatment of polyps, tumors, and growths while detoxifying the meridians and chakras.

Green Fluorite

Green fluorite stabilizes growths and chaotic energies.

Peridot

Peridot regenerates and heals the skin while acting as a tonic that strengthens and purifies the body.

Seraphinite

Seraphinite powerfully regenerates and acts against the growth and reproduction of cancer cells.

BLUE STONES

Blue stones are also important, for they calm overactive and chaotic growth, encouraging homeostasis.

Lapis Lazuli

Lapis lazuli regenerates the body, encouraging healing and the liberation of repressed memories and emotional baggage.

OTHER STONES

Prophecy Stone

Prophecy stone supports during chemotherapy and radiotherapy.

Selenite

Selenite stimulates cellular regeneration, offering preventative measures while repairing damage caused by cancer.

Sugilite

Sugilite guides toward the discovery of the profound and subtle causes of a cancer while emitting beneficial energies that favor recovery.

Contraindications

It is recommended to avoid stones and crystals with stimulating colors, such as yellow, orange, and red. Clear quartz should also be avoided as it is highly stimulating.

Stone Placement

The placement of your stones will depend on the location and type of cancer, although it is important to treat the solar plexus, the center of our emotional life, regardless of the particulars of the diagnosis. An appropriate stone can be placed at the plexus for intervals of between five and fifteen minutes, depending on the sensitivity of the child and the quality, size, and power of the stone being used. However, an appropriate stone can be worn close to the afflicted area, held in the hands, or worn as jewelry for extended periods of time.

Children have also been known to respond well to stone elixirs. A few drops of a good-quality elixir can be rubbed onto the solar plexus, under the feet or on the hands, or put in some water for the child to drink, two or three times per day.

◆ ◆ ◆

THE GRIEVING PROCESS

NANCY BÉDARD

Like adults, children will one day be confronted with death and significant losses. These inevitable trials and tribulations are part of the cycle of life and help shape and change any individual, no matter how young they may be. Each child expresses the sorrow and feelings that are triggered by such a loss in his or her own way. Although children react differently than adults, their suffering, which is often underestimated, is just as profound.

When they lose someone close to them, children, who are in the middle of their development process, have as great a need as adults to be listened to and understood when they express their emotions. The attitude of their parents and those around them is crucial. How can we respond to their distress when we ourselves are distraught? Frankness and honesty are uppermost on the list.

Because of their particular life experience, all children have their own ideas about death, be they right or wrong. Evading the truth and trying to distract them from their sorrow can do more harm than good. Explaining things simply, in age-appropriate language, helps build solid foundations that will enable them to understand and accept death. Including children in the grieving process by letting them participate in funeral rituals and sharing our own feelings show them that although the situation is painful, it is also natural. In this way, we can prevent them from feeling left out and encourage them to express what they feel.

When someone dies, those involved are at a loss themselves, so it is hard for them to provide children with the comfort and stability they need. A serious loss sparks physical, psychological, emotional, and spiritual distress. Stones and crystals can play a role on all these levels and offer additional support that will help children come to terms with what has happened. They emit an energy that words alone cannot and transmit the message that there is more than the physical body, inspiring confidence in the divine wisdom of each soul.

Very Young Children

When they are very young, children don't know what death means. They experience the loss as an absence or a separation. They wonder where the person who looked after them so well has gone. What is happening to those who love them? Attuned to the energy of everyone around them, they feel adults' helplessness and distress, which in turn affects their emotional stability and behavior.

SUGGESTED CRYSTALS

Celestite, Rose Quartz

Celestite and rose quartz are ideal supports for very small children. Celestite, which bestows maternal energy, envelops children in protective love, partly compensating for the absence of the deceased parent or the lack of attention and care from those around them, who are too devastated by their loss to respond appropriately. Celestite will also help them adjust to the new situation.

The gentle color of rose quartz brings calm and tenderness to their environment, reassuring them and reminding them that they are loved. It thus soothes them by reducing their anger and fears. The quartz of unconditional love, rose quartz comforts children overwhelmed by sorrow at this stressful time.

Ages Three to Five

At this age, children still live in a world of make-believe. They don't clearly understand what death means, but they think they can bring the person back to life as they do in their games. They may even imagine that they are responsible for this death, which can generate anxiety and the fear that someone else close to them may also die. Some behaviors, such as skills regression, bed-wetting, thumb sucking, or talking baby talk, are obvious indications that the child is affected by the loss.

SUGGESTED STONE AND CRYSTALS

Ametrine

Ametrine, which combines the properties of amethyst and citrine, helps dispel fears and anxiety, bringing harmony and inner peace. It inspires the joy in living needed to balance the emotional body, which enables children to continue to live their childhood to the fullest.

Aventurine, Rose Quartz

Aventurine and rose quartz are recommended for this age group. Linked to the heart chakra, they calm the intense emotions raised by the loss, imparting love and tranquility to alleviate the feeling of abandonment. They also help reassure children of the place they occupy within the family. As well, these stones transmit the confidence that they can again love and be loved and build emotional ties with someone else without fear of loss.

Ages Six to Twelve

Children in this age bracket have a better understanding of what death means. They understand that the person will not return. Sometimes believing that only old people die, they can be devastated by the death of a young person and even transpose the possibility of dying to themselves. This in turn could contribute to the development of fear in general and fear of death in particular.

SUGGESTED STONE AND CRYSTALS

Amethyst

By sustaining emotional balance, amethyst can help children cope with the family disruption that has such a strong impact at this age. It dispels fears and calms anxiety, supporting children so that they can adapt to the changes brought about by the loss. Amethyst is also known for strengthening the immune system, which could be destabilized at this time.

Citrine

Citrine restores good humor and joy by attenuating the worry and anger grief brings. At this age, children perceive and interpret adult silences, which can cause them to lose confidence and fear the future. Citrine confers reassurance and courage and supports them in expressing their feelings and needs.

Fluorite

Fluorite is also very useful in helping to maintain concentration and the capacity to learn at school during the grieving period. It brings balance to the mind, clarifies ideas, and inspires self-confidence, thus dispelling confusion. It enhances the memory and promotes understanding, two important factors in academic achievement.

Rose Quartz

When children are deeply wounded by the loss of a loved one, rose quartz can help them absorb the emotional shock of this bereavement. It eases their suffering, enabling the reaffirmation of love and safeguarding them from depression.

Adolescents

Unlike their younger siblings, adolescents are able to understand death and its implications. However, they often react differently from adults. As they are in transition between childhood and the adult world, their feelings are intense, which at times can lead to behaviors that are incomprehensible to those around them. The death of a friend can affect them as much as the death of a parent. Close to their emotions and receptive to their inner feelings, teenagers are often tempted to repress them just because they are so invasive. They then put on a show of indifference to mask their pain.

SUGGESTED STONES AND CRYSTAL

Amethyst

A period of mourning can be an extremely difficult time for a teenager. During this important period of transformation, it isn't always easy to understand them or clearly interpret their feelings. Amethyst can support them in their ordeal, clarifying their thoughts and helping them remain focused and in control of their faculties. It can thus prevent them from falling into the trap of drug, alcohol, or even sex addiction to numb their suffering.

Kunzite, Rhodonite

Pink stones like kunzite and rhodonite also confer the beneficial energy of love so necessary for absorbing any emotional shock and thus prevent panic attacks. These stones transmit the wisdom needed to trigger forgiveness and diminish anger.

Kyanite

Kyanite, a stone of tranquility, unites the heart and mind, drawing light down to enable us to observe the different aspects of the situation from which the attachment and suffering stem. Cutting through fears and obstacles, it encourages the assertion of truth. It thus enables adolescents to express themselves and communicate their feelings, effectively calming the uncertainties sparked by what remains unspoken. Kyanite is useful for both adolescents and their parents because it can help them understand each other and communicate.

Although this chapter focuses on the grieving process among children, particularly when someone close to them has died, it can be applied to any loss. Parents' divorce, a move, changing schools, and a parental attitude that doesn't meet a child's expectations are all perturbing events that impose change. That's why it is extremely important for children and teens to receive help at these troubled times. Support from an adult is crucial, of course, but the free-flowing sincere support of stones and crystals is also beneficial. They are steadfast allies that will help us adapt to the changes wrought by loss.

◆ ◆ ◆

An A-Z of Suggested Stones

Agate

Amazonite

Amber

Amethyst

Amethyst (cluster)

Ametrine

Andradite garnet

Apache tear obsid-
ian

Apatite

Apophyllite

Aqua aura quartz

Aquamarine

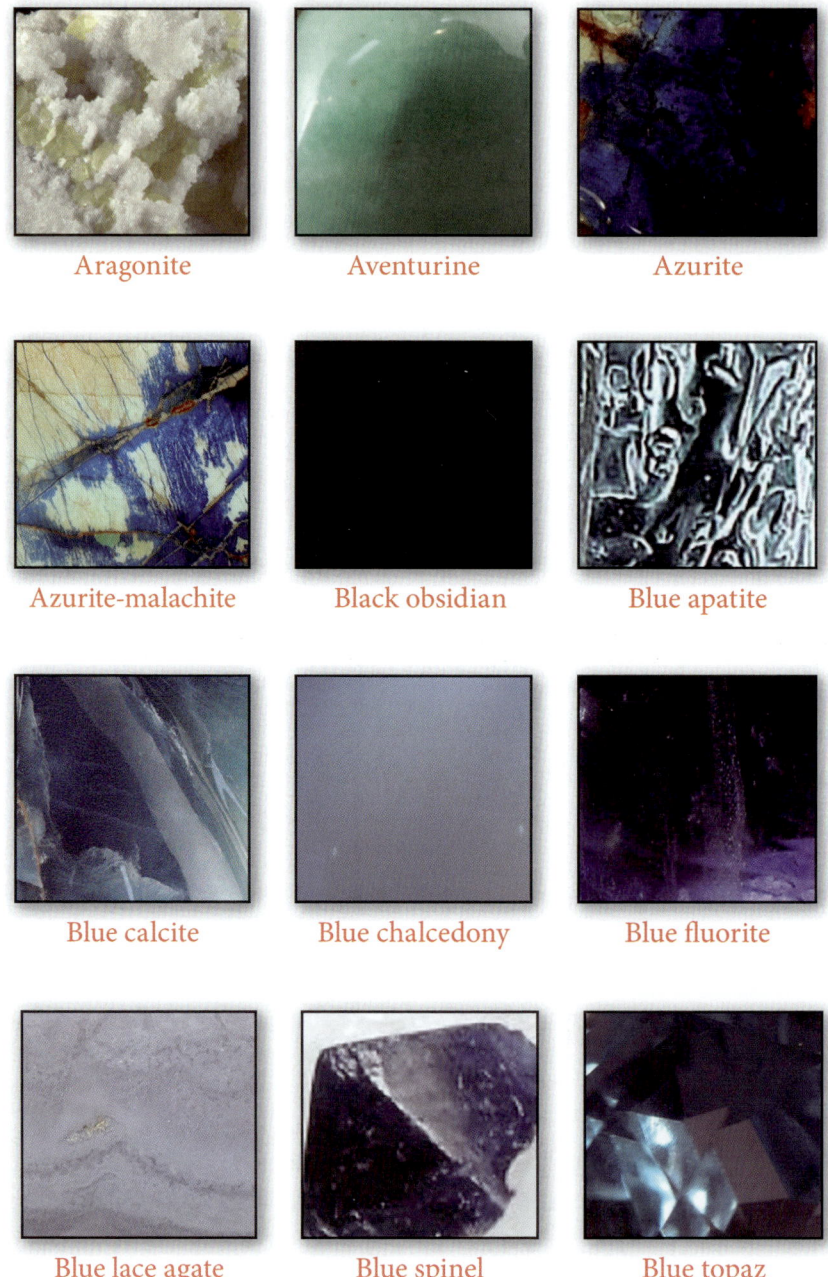

Aragonite

Aventurine

Azurite

Azurite-malachite

Black obsidian

Blue apatite

Blue calcite

Blue chalcedony

Blue fluorite

Blue lace agate

Blue spinel

Blue topaz

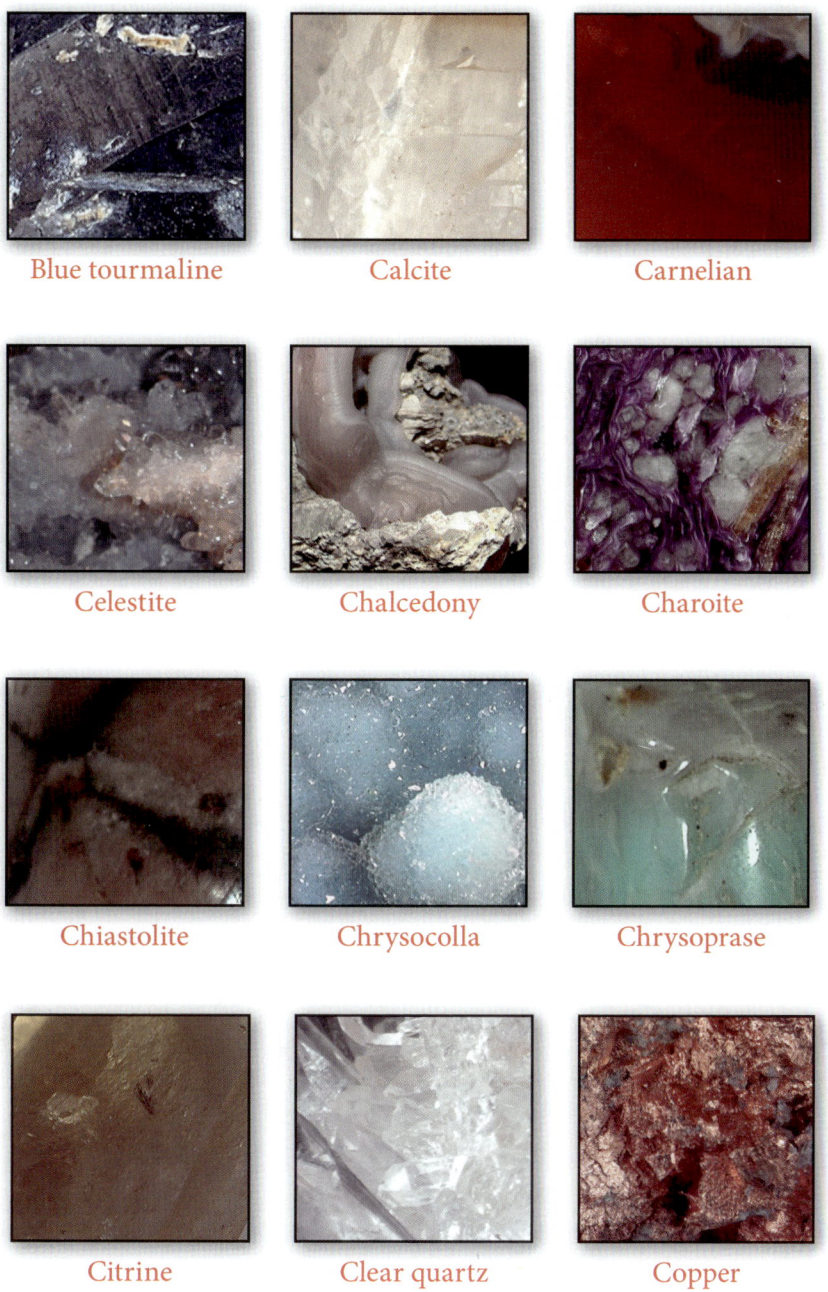

Blue tourmaline

Calcite

Carnelian

Celestite

Chalcedony

Charoite

Chiastolite

Chrysocolla

Chrysoprase

Citrine

Clear quartz

Copper

Danburite

Dioptase

Emerald

Epidote

Fluorite

Garnet

Green apatite

Green calcite

Green fluorite

Hematite

Herkimer quartz

Jade

Jasper

Kunzite

Kyanite

Labradorite

Lapis lazuli

Larimar

Lepidolite

Magnesite

Magnetite

Malachite

Moldavite

Moonstone

Moss agate

Nephrite jade

Orange calcite

Peridot

Petrified wood

Pink calcite

Pink tourmaline

Prehnite

Prophecy stone

Purple fluorite

Pyrite

Rhodochrosite

Rhodonite

Rose quartz

Ruby

Rutilated quartz

Rutile

Selenite

Selenite (cluster)

Seraphinite

Serpentine

Smithsonite

Sodalite

Sugilite

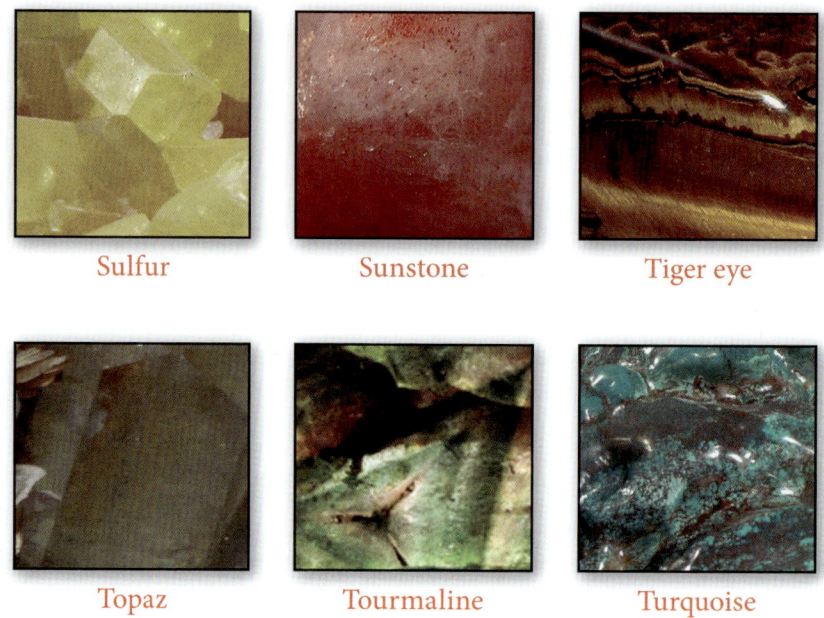

Sulfur   Sunstone   Tiger eye

Topaz   Tourmaline   Turquoise

Suggested Stones and Crystals by Subject

Adoption
Amethyst
Citrine
Emerald
Rose quartz

Allergies
Amber
Amethyst
Apophyllite
Aquamarine
Aventurine
Citrine
Danburite
Emerald
Lepidolite
Rhodochrosite
Rutilated quartz

Anorexia
Lepidolite
Rose quartz
Sunstone
Topaz

Anxiety (First Day of School)
Amethyst
Aquamarine
Charoite
Citrine
Labradorite
Lapis lazuli
Larimar
Rose quartz
Ruby
Sodalite
Sugilite

Asthma
Apophyllite
Magnetite

Malachite
Pyrite
Rhodochrosite
Rutilated quartz

Autism
Apatite
Blue lace agate
Charoite
Lepidolite
Sodalite
Sugilite

Bed-wetting (Enuresis)
Amazonite
Chrysoprase
Citrine
Nephrite jade
Peridot
Prehnite
Rose quartz
Sunstone

Bullying at School
Ametrine
Apatite
Aquamarine
Aventurine
Citrine
Dioptase
Garnet
Kunzite
Rose quartz
Ruby
Tiger eye

Childhood Cancer
Aventurine
Green calcite
Green fluorite
Lapis lazuli
Malachite
Peridot

Prophecy stone
Selenite
Seraphinite
Sugilite

COLIC
Magnesite
Malachite
Serpentine
Turquoise

CONCENTRATION
Carnelian
Citrine
Fluorite
Lepidolite
Pyrite
Sodalite

CONFLICTS
Aventurine
Epidote
Jasper
Labradorite
Rhodonite
Sugilite

DIGESTION
Amber
Chrysocolla
Chrysoprase
Citrine

DYING CHILD
Amethyst
Aquamarine
Carnelian
Charoite
Danburite
Kunzite
Kyanite
Pink calcite
Rhodochrosite

Rhodonite
Rose quartz
Sugilite

DYSLEXIA

Apatite
Carnelian
Fluorite
Kyanite
Petrified wood
Sugilite
Tourmaline

ECZEMA

Amber
Aventurine
Sulfur

FEARS

Agate
Amethyst
Aventurine
Carnelian
Celestite
Citrine
Herkimer quartz
Kunzite
Lepidolite
Orange calcite
Rose quartz
Sodalite
Sugilite
Tiger eye

FEVER

Amber
Green calcite
Hematite
Sodalite
Sulfur

GRIEVING PROCESS
Amethyst
Ametrine
Aventurine
Celestite
Citrine
Fluorite
Kunzite
Kyanite
Rhodonite
Rose quartz

GROUP INFLUENCE
Agate
Aquamarine
Aventurine
Citrine
Fluorite
Herkimer quartz
Labradorite
Lapis lazuli
Lepidolite
Ruby
Sodalite
Sugilite
Tiger eye

HYPERACTIVITY
Agate
Amazonite
Amber
Ametrine
Apatite
Carnelian
Citrine
Fluorite
Jasper
Kyanite
Petrified wood
Pyrite
Rhodochrosite
Rhodonite
Rose quartz
Tiger eye

Immune System
Amber
Amethyst
Ametrine
Aqua aura quartz
Aquamarine
Aragonite
Calcite
Chalcedony
Chiastolite
Emerald
Fluorite
Lepidolite

Meditation
Celestite
Citrine
Fluorite
Kyanite
Rose quartz

Menstrual Cramps
Magnesite
Malachite
Moonstone
Serpentine

New Baby
Aventurine
Carnelian
Celestite
Dioptase
Lepidolite
Malachite
Moonstone
Pink tourmaline
Pyrite
Rose quartz
Smithsonite
Sunstone

NIGHTTIME FEARS
Aquamarine
Amethyst
Celestite
Charoite
Kunzite
Rose quartz

ORAL PRESENTATIONS
Aquamarine
Blue calcite
Blue lace agate
Lapis lazuli
Turquoise

ORGANIZATION
Fluorite
Jasper
Petrified wood
Pyrite
Sodalite

PHYSICAL TRAUMA
Amazonite
Amber
Amethyst
Andradite garnet
Apophyllite
Aquamarine
Aragonite
Aventurine
Azurite
Azurite-malachite
Blue chalcedony
Blue fluorite
Blue lace agate
Blue spinel
Blue tourmaline
Calcite
Charoite
Chrysocolla
Citrine
Copper

Dioptase
Emerald
Fluorite
Green apatite
Green calcite
Hematite
Jade
Kyanite
Labradorite
Lapis lazuli
Larimar
Magnesite
Magnetite
Malachite
Moss agate
Pyrite
Rhodochrosite
Rhodonite
Rose quartz
Rutile
Selenite
Sodalite
Sunstone
Turquoise

Preparing for Exams

Chrysoprase
Citrine
Fluorite
Kyanite
Orange calcite
Pyrite
Rhodochrosite
Rhodonite
Smithsonite
Sodalite

Rejection

Apache tear obsidian
Black obsidian
Moldavite
Rhodochrosite
Rhodonite
Rose quartz

SELF-ASSERTIVENESS
Amazonite
Aquamarine
Blue calcite
Blue lace agate
Blue topaz
Citrine
Kunzite
Kyanite
Lapis lazuli
Malachite
Pink calcite
Rose quartz

SPEECH
Chalcedony
Citrine
Lapis lazuli

STOMACHACHE
Amazonite
Amethyst
Amethyst or selenite cluster
Aqua aura quartz
Aquamarine
Blue apatite
Calcite
Chrysoprase
Rose quartz
Selenite

STRESS AND ANXIETY
Amethyst
Aquamarine
Fluorite (blue, purple or green)
Kunzite
Lapis lazuli
Rhodonite

TEETH AND TEETHING
Amber
Amethyst
Apatite

Aquamarine
Aventurine
Calcite
Celestite
Clear quartz
Fluorite
Garnet
Green calcite
Labradorite
Magnesite
Magnetite
Malachite
Rose quartz
Selenite
Tiger eye
Turquoise

URINARY TRACT INFECTIONS

Amber
Aventurine
Carnelian
Emerald
Hematite
Malachite
Prehnite

BIBLIOGRAPHY

Barbour, Chandler, Nita H. Barbour, and Patricia A. Scully. *Families, Schools, and Communities: Building Partnership for Educating Children.* New York: Prentice Hall, 2008. http://www.education.com/reference/article/peer-group-influence/?page=2.

Bourcier, Sylvie. *L'influence des pairs au sein d'un groupe d'enfants.* http://www.aveclenfant.com/index.php?option=com_content&view=article&id=194

Dewar, Gwen. *Nighttime Fears in Children: A Guide for the Science-minded Parent.* 2008. http://www.parentingscience.com/nighttime-fears.html.

Gienger, Michael. *Crystal Power, Crystal Healing: The Complete Handbook.* New York: Sterling Publishing, 2007.

—. *Manuel de lithothérapie, ou l'art de se soigner avec les pierres.* Paris: Éditions Véga, 2005.

The Group of 5. *Crystals and Stones: A Complete Guide to Their Healing Properties.* Berkeley, CA: North Atlantic Books, 2010.

—. *The Eight Crystal Alliances: The Influence of Stones on the Personality.* Montreal: Paume de Saint-Germain Publishing and Berkeley, CA: North Atlantic Books, 2010.

—. *L'Influence des Pierres: Une approche psychologique.* Montreal: Éditions Paume de Saint-Germain, 2010.

—. *Manuel de lithothérapie: Apprenez à mieux connaître le royaume des pierres et des cristaux: Thème 1: Qui sont ils?* Montreal: Éditions Paume de Saint-Germain, 2008.

Guidou, Nadège. *L'enfant et le deuil, comment les plus petits comprennent-ils la mort?* http://assmat33.forumactif.com/t69-l-enfant-et-le-deuil.

Hadley, Josie, and Carole Staudacher. *Hypnosis for Change.* Oakland, CA: New Harbinger, 1996.

Hall, Judy. *The Crystal Bible.* Cincinnati: Walking Stick Press, 2003.

Kübler-Ross, Elisabeth. *On Children and Death.* New York: Macmillan, 1983.

Laplante, Francine. *Jusqu'au bout de ta courte vie: L'histoire de six jeunes héros et leurs parents endeuillés.* Montreal: Les Éditions La Presse, 2008.

Les Coopératives Funéraires du Québec. *La Gentiane: Le deuil chez l'enfant.* http://lagentiane.org/1_2_7.htm

Naudin, Claude, and Nicole Grumbach. *Larousse medical.* Paris: Larousse, 1995.

Ouellet, Hélène. *Le deuil chez l'enfant et l'adolescent. Prévenir et agir.* Quebec: Santé et Services sociaux, 2009. http://publications. msss.gouv.qc.ca/acrobat/f/documentation/2009/09-235-12F_08.pdf

Paulin, Jean-Yves. *La Mystique des Pierres.* Lyon: Éditions du Cosmogone, 2002.

Pincus, Donna, and John Otis. *Fears, Phobias and Anxiety.* The Child Anxiety Network, 2001. http://www.childanxiety.net/Fears_Phobias_Anxiety.htm.

Simmons, Robert, and Naisha Ahsian. *The Book of Stones: Who They Are and What They Teach.* Berkeley, CA: North Atlantic Books, 2007.

Sri Adi Dadi, *Jyoti for Kids: A Meditative Technique of Purification by the Light* (CD-ROM included). Montreal: Paume de Saint-Germain Publishing, 2009.

Vachon, Daniel. *Le deuil chez l'enfant.* Commission scolaire des Hauts-Cantons. http://www.aqps.qc.ca/public/publications/bulletin/15/15-2-05.html.

INDEX

WHO IS THE GROUP OF 5?

Klaire D. Roy, Lisa C. Bergeron, Kristiane Roy, Jacqueline D. Sylvain, Ginette Tétreault

We are passionate therapists who, each in turn, collaborate to write various texts on therapy with stones. We all agreed on the name The Group of 5 for the publication of our first book, titled *Crystals and Stones: A Complete Guide to their Healing Properties*, published in 2008. But what does this name signify for us?

Choosing our name, The Group of 5, was important because it had to represent what form we would adopt in the literary world. Accordingly, the name was chosen in relation to the pentagram, the five-pointed star, representing the ideal human being drawn and conceived by Leonardo da Vinci. This person symbolizes the accomplished, active, and productive aspect within us all.

Due to its calligraphic aspect, the pentagram also relates to writing. It features five graphic elements, which we extrapolate to the five senses that assist us in composing our articles, constructing chapters for our books, and mutually sharing our knowledge. Moreover, we exist in a world defined by the five elements: the earth, from which crystals evolve; air (oxygen), which makes life possible and of which several minerals are composed; water, which erodes and shapes stones while at times infiltrating them; fire, which heats and aids in the transformation of minerals; and the ether, which plays a mysterious role in the transmutation of elements.

This star, a sign of good fortune, is, if you look closely, made up of five interlaced A's, the first letter of the Latin alphabet and a symbol of academic excellence. We hope that this star illuminates, as much as possible, those who want to know more about the world of stones and crystals.

The number 5 also signals concrete intelligence resonating with science, which ensures that the proposed hypotheses are correct and not merely based on intuition. We note as well that it is related to the human form: five fingers, five extremities, five senses, and five tastes. Formed by a semicircle and a half square, it also expresses the polarities, yin (circle) and yang (square) united in a single character. The expression "five by five," based on the idiom for an ideal radio transmission, symbolizes our desire to simplify all while honoring the mineral world so that all can understand its nature.

Each text proposed by one of the members of The Group of 5, whether for a chapter in a book or an article, has been corroborated by the other team members, who study it and support the author's work. Teamwork is essential for us, as the mineral world guards so many secrets that we believe it is next to impossible that they will all be revealed some day. Beyond our notions of stones and crystals, we share a joy and passion that invigorates each of our encounters.

Profits derived from book sales are reallocated for the publication of new books dealing with stones and crystals, and to support a project for the creation of a therapy center in Potton Springs, situated in the Eastern Townships of Quebec, Canada. We take our profession to heart and are truly grateful to the mineral world for letting us work in such close collaboration with it. It is a world of treasures and discoveries that we want to share with all those who would like to experience or better understand the mineral world. Are we not also constituted of minerals?

To learn more, or simply to share new ideas or queries, contact us by email at info@IMBM.ca. We welcome any and all new ideas, suggestions, and questions, so don't hesitate to write to us or read our works. Your comments are precious and appreciated by us all.

Klaire D. Roy
Director of the Medicine Buddha Mandala Institute and The Group of 5

◆ ◆ ◆

CONTRIBUTORS

Nancy Bédard
Author, speaker, and businesswoman

Laëtitia Betton
Lithotherapist, naturopath

Lynda Blanchette
Social worker

Gaëtan A. Brouillard
Physician

Bertrand Corbeil
Lithotherapist, reflexologist, massage therapist

Annie Dufresne
Teacher

Lise Dussault
Lithotherapist

Joani Gagnon
Massage therapist, physiotherapist

Johanne Marier
Lithotherapist

Elianne Meier
Teacher

François Nicol
Lithotherapist

Eliane A. Panneton
Lithotherapist, naturopath

Philippe Soreau, P. Eng.
Lithotherapist

201

01 14